Praise for NEUROSELL

"Whether you're a sales person or a sales leader, your time is too valuable to read another sales book that tells you what you already know. *NeuroSelling*® is very different. It will fundamentally change the way you think about selling and if you apply the *NeuroSelling*® process, it will dramatically improve your results."

—MATT RIEBEL, CHIEF SALES OFFICER, PROTECTIVE LIFE

"This isn't just another book on mastering sales acumen. *NeuroSelling*® is a real-world guide on building trust, strengthening relationships and driving genuine and intelligent sales conversations."

—BRAD LARSEN, SENIOR DIRECTOR,
BUSINESS DEVELOPMENT, MITSUBISHI ELECTRIC

"Understanding your customer, how they think and what motivates them to say 'yes' to your sales team is about more than just a relationship—it's an artful science that Jeff explains with masterful simplicity. His insights, experience and methodology in *NeuroSelling*® is a pathway to profitability and should be required reading for any team serious about excelling in today's dynamic marketplace."

—JIM PEARSON, CEO, NICO CORPORATION

"This book goes straight to the heart of why traditional sales techniques don't give us the results we want. Then it provides a systematic, scientifically proven path from the inside out; a true people-first approach. Thanks, Jeff, for showing us a better way!"

—NORA STEWART, CEO, ENTHEOS

"Finally! A sales book that dismantles all of the 'slimy' sales gimmicks of old and addresses the heart and science of a purposeful and successful sales conversation. Do not engage another customer until you have read this book."

—TOMMY SEAY, DIRECTOR, BUSINESS DEVELOPMENT, AMERICAN EXPRESS

"It's rare to find someone who understands the art AND science of sales so well. Jeff Bloomfield is at the top of the list!"

—EDDIE YOUNG, SR. VICE PRESIDENT, SALES, FRESHPET

"I am a skeptic by nature. I put half my sales team through *NeuroSelling*® and the other half did things the old way. The result? The *NeuroSelling*® group increased sales by 35 percent over their peers. Naturally, everyone got the book after that! Jeff is a master communicator. This is the most effective methodology I have ever seen!"

—ANDREW WYANT, PRESIDENT, ISSA

"*NeuroSelling*® provides an understanding into the science behind customer decision making and provides a road map to master it. Most sales books are either too theoretical, too basic or another version of a similar methodology but with a different acronym. *NeuroSelling*® is a fantastic balance of 'why' and 'how.' The book is in and of itself, self-evident, as the stories in it compel you to read on. I highly recommend reading it and applying its learnings!"

—CHAD BRINES, VICE PRESIDENT, SALES, OCULAR THERAPEUTICS

"As a CEO, I was engaged from the first page! I found this content so relatable and applicable. I really eat this type of knowledge up. Jeff's insight into more impactful communication will continue to help me refine my skills as a leader. Bravo!"

—JARROD MCCARROLL, CEO, WEBER, INC.

"Understanding the science behind customer decision making and having a road map to master it? Brilliant."

—DAVE NURRE, REGIONAL PRESIDENT, USI INSURANCE

"Jeff teaches you how to build trust faster and drive change through customer conversation with a unique and science-based approach. He is a product of his product in that, he gives the lessons by using the very concepts to train you on their value! This is a must-read for any sales professional. With *NeuroSelling*®, Jeff helps you develop the skills and principles necessary for anyone to build a more satisfying and fulfilling sales career."

—Tom Ziglar, CEO, Ziglar, Inc.

"A must read if your sales target is a living, breathing, thinking human being! These principles seem so obvious after reading the book, but how many sales reps actually use them in their day to day? Great read, it will change the way you sell!"

—Joe Schowalter, Global Marketing Director, Ethicon, Inc.

"Jeff is a master communicator and this book explains why! A must read for your sales team, and anyone wanting to improve their connections with people—*NeuroSelling*® is the perfect balance of 'why' and 'how.'"

—Caitlin Clipp, Executive Director, United Healthcare

"I've worked with clients in the Fortune 100 who haven't discovered what Jeff Bloomfield has taught for years: B2B sales success is a system. In *NeuroSelling*®, Jeff provides a working understanding of the science behind sales, then shows how to use it to positively nail that first critical customer conversation. I expect to see his book soon on executives' desks everywhere."

—Matthew Pollard, Author of *The Introvert's Edge: How the Quiet and Shy Can Outsell Anyone*

NEURO
SELLING
2.0

Mastering the Customer Conversation
Using the Surprising Science of Decision-Making

JEFF BLOOMFIELD
DAN DOCHERTY, PHD

AP Axon Publishing

NeuroSelling® 2.0: Mastering the Customer Conversation
Using the Surprising Science of Decision-Making

For information about this title or to order other books and/or electronic media, contact the publisher:
Axon Publishing, LLC
info@braintrustgrowth.com

Revised edition 2025

ISBN: 979-8-9990907-0-6 (Hardback)
978-1-7337870-6-2 (Paperback)
978-1-7337870-7-9 (eBook)

Printed in the United States of America

Cover and Interior design: 1106 Design

*"The first and simplest emotion
which we discover in the human mind, is curiosity."*

—Edmund Burke (1729–1797), Irish Statesman,
Author, and Philosopher

CONTENTS

INTRODUCTION TO NEUROSELLING 2.0

WHEN WE PUBLISHED the first edition of NeuroSelling, we knew we were onto something powerful. The response was overwhelming—sales professionals from around the globe began implementing these neuroscience-backed principles and achieving remarkable results. But we're not ones to rest on past success. As my team and I traveled the world speaking and working with organizations of all sizes, we continued to observe, research, and refine. The result is what you now hold in your hands.

NeuroSelling 2.0 isn't just an update—it's a complete reimagining of what's possible when you truly align your sales approach with how the human brain actually makes decisions.

Since the original publication, the world of selling has transformed dramatically. The digital revolution accelerated by seven years during the pandemic. Artificial intelligence has moved from science fiction to everyday sales tool. The neuroscience of decision-making has advanced with breakthrough discoveries about how our brains respond in virtual environments versus face-to-face interactions.

But perhaps most significantly, we've had the privilege of witnessing thousands of sales professionals apply these principles across industries, cultures, and sales contexts. Their experiences, challenges, and triumphs have profoundly shaped this new edition.

We've incorporated extensive new research validating the neurological foundations of effective selling. We've expanded the implementation frameworks with step-by-step guidance for digital, hybrid, and AI-enhanced selling environments. The book now includes detailed case studies documenting the measurable impact of NeuroSelling in organizations, from midsize companies to global enterprises across pharmaceuticals, manufacturing, financial services, and technology.

What can you expect from NeuroSelling 2.0? Beyond mastering the science of human decision-making, you'll gain practical skills that directly impact your results.

You'll learn how to establish deep personal trust in seconds—even through digital channels—using evidence-based trust-acceleration techniques. You'll discover how to craft the five essential narratives that drive urgency to change, calibrated for your specific customer contexts. You'll develop the ability to anticipate and remove barriers to change before they arise, using the latest research on psychological resistance. And you'll understand how to leverage AI as a powerful complement to—never a replacement for—authentic human connection.

Our clients consistently report 30–50 percent increases in closing ratios, 25–40 percent reductions in sales cycles, and significant improvements in average deal size—all while strengthening customer relationships and reducing the stress of selling.

But the impact goes beyond numbers. The most rewarding feedback I receive comes from sales professionals who tell me, "NeuroSelling hasn't just transformed how I sell—it's transformed how I communicate in every area of my life."

That's the true power of this approach. When you understand the science of human connection and decision-making, you don't just become a better salesperson. You become a more effective communicator, a more empathetic listener, and a more influential leader.

In the words of one client: "I used to see myself as just another sales rep pushing products. Now I see myself as a trusted advisor who genuinely helps people solve meaningful problems." That transformation—from transactional seller to trusted problem-solver—is what NeuroSelling 2.0 is all about.

Our beliefs drive our behaviors. Our behaviors consistently form our habits. Our habits are what actually generate our results in life. If you want different results, yes, there are some habits that will need to change, but as you read this book, I encourage you to ask yourself, "What do I believe?" It is there that you will find the spark that leads to sustainable change in the way you communicate.

The pages that follow may challenge what you think you know about selling. They'll ask you to leave your comfort zone and try approaches that might feel unfamiliar at first. But I promise you this: If you embrace these principles, practice these techniques, and persist through the initial learning curve, you'll never look at sales conversations the same way again.

Are you ready to transform how you connect, communicate, and create change? Let's begin.

WHY THEY BUY (AND WHEN THEY DON'T)

"The really important question is this: Why would a person do business with you at all?"

—ROY H. WILLIAMS, *THE WIZARD OF ADS*

MY SON, DREW, has a life-threatening allergy. Ingesting the smallest amount—or simply being exposed to peanuts—can trigger anaphylaxis. It can be deadly for him, depending on how much he's exposed to and how long it takes to treat it with epinephrine.

Our family has EpiPens stashed everywhere. Drew's backpack, my wife's purse, a sling pack, cargo shorts pockets, vehicles, at home, at church, Drew's taekwondo bag—anywhere we're going to be, there's going to be an EpiPen within reach.

Before the beginning of every school year, we meet with Drew's teachers and school staff to emphasize the severity of his allergy and review their plan for keeping our son safe and healthy. Year after year since kindergarten, we would begin the frustrating ritual of scheduling the meeting with his school staff and administrators. We had struggled through kindergarten, then first grade, followed by second, and each year, it became more and more frustrating at the seemingly passive, almost dismissive attitude we received from the school staff. In our eyes, it was as if they didn't really view this as a problem at all. Meanwhile, we were seeing article after article of mourning parents whose child died from unnecessary peanut exposure due to surroundings (including schools) that didn't have the necessary protection in place for their child.

For the beginning of his third-grade year, we walked into the room where the teachers and principal had assembled for our yearly ritual. Once again, we could see that the teachers and staff were not viewing this as importantly as we knew the problem to be. New year, same perceived lack of urgency. From their perspective, I'm sure they felt they were used to dealing with parents whose children had food allergies and that this was a run-of-the-mill conversation for them. They were nice and attentive, but their focus seemed to be on reassuring us, the scared parents, that they saw this kind of thing all the time and would take care of Drew "just like we have all the other students who've come through our school."

My wife went through her annual presentation of all the signs and symptoms of anaphylaxis with all the supporting facts and data to support the need for urgent attention and action. The response was the usual: "Yep. We got it." "We will make sure none of the kids with peanut butter or peanut products gives any to Drew." "We will do our best to be sure he doesn't eat the wrong thing, etc. etc. etc." I was frustrated. My wife was close to tears. We weren't exaggerating or being dramatic when we said "life-threatening." These teachers just saw two scared parents. We saw a group of people who we believed didn't understand how serious Drew's allergy was. In other words, they didn't see a problem. They were just dealing with nervous parents; we were fighting for our son's life.

We needed them to change. We needed them to see the need to move from their current "status quo" to a place that was actually mutually beneficial for both parties.

What could we do to get them to get it? To really, truly *get it*?

Apparently at an impasse, we all fell silent for a moment while we gathered our thoughts and tried to figure out a way forward. I was racking my brain and trying my best to suppress the stress and frustration I could feel coursing through my veins. I help people communicate with prospects in a way that drives urgency to change every day! Yet I was momentarily stuck. My way forward in this moment was likely to include a scorched-earth communication approach, but I was fervently holding my tongue. With tears in her eyes, my insightful and wisdom-filled wife, Hazel, looked up from the undersized elementary school table and quietly asked, "Would any one of you let a student bring a loaded gun to school?"

Immediate confusion. I could almost read the teachers' thoughts: *Wha-huh? Weren't we just talking about peanut butter*?!

Hazel pressed on: "Every time you let a student bring anything made from peanuts into the classroom, it's the same as if they're pointing a loaded gun at Drew. That's how dangerous his allergy is."

Yes, I understand that in today's culture, using a gun analogy, particularly in a classroom setting, may make some of you a tad bit uncomfortable. But you must know my wife. She is humble and kind. She is naturally a reserved person who isn't prone to emotional language or hyperbole. In addition, she was a former educator and everyone in that room knew her and trusted her. As those words rolled from her lips, in that one moment, something shifted. Something changed.

All the information and conversation up to that point appeared to finally click. The attitude of everyone in the room went from politely patronizing a couple of overprotective parents to complete buy-in and cooperation. They finally connected with our message. They finally had an urgency to change. They had a different perspective on the problem.

Whether you realize it or not, the issues we faced in that elementary classroom sitting on those multicolored chairs a couple of sizes too small for adults are exactly the same thing you face in your day-to-day sales conversations.

The Great Disconnect—Why Your Message May Miss the Mark

Why wasn't providing the teachers and administrators the information, the facts, the data, and the signs and symptoms around Drew's life-threatening allergy enough? What was it that my wife said that finally clicked for everyone in that room?

The epiphany certainly didn't come from Hazel's tears or my obvious frustration. Schools deal with tearful moms and angry dads just like us all the time. While our educators are not unfeeling robots, after so many meetings with unhappy parents, they have to become thick-skinned during these types of meetings just to keep their own sanity. We weren't dealing with uncaring drones—we just couldn't get past the professional outer shell they'd created after so many meetings with helicopter parents who really did overreact to every minor issue, and quite honestly, they felt they had the answers, knew enough, and didn't need further "convincing" to change their actions or behaviors.

And aren't your customers the same way?

When our clients' salespeople snag a few minutes of their customer's time, be it a surgeon standing outside the operating room, an engineer in his office, or walking into a conference room to pitch to the customer's executive team, they face the same situation my wife and I did—a roomful of people who already have their guard up, who already assume they know what the salesperson is going to say, and who already believe they are likely doing things well enough to not need whatever they're selling.

You know how this scene plays out, don't you?

Here's the frustrating thing: You know with certainty that your product is superior to the competition's. If you're like most of our clients, you probably charge a premium of 10–20 percent or more over the competition because of how superior your product or service is.

Despite you knowing and even stating that, the person on the other side of this conversation has heard it all before. Every salesperson claims to have the best, be the best, and care the best. Every one of them. They see you as just another anonymous parent—excuse me, sales rep—singing the same tune they've heard so many times.

You know you're different.

But how can you get them to *get it*?

My wife uttered three sentences that changed the entire feel of the room and transformed those educators into partners and guardians of our child. I felt elated. But when I later replayed the scene in my head, I could have kicked myself. Of course, what she did worked—in a microcosm, it was exactly how we teach people to successfully sell every day! They trusted her personally (connection) and they believed she knew her stuff (credibility), so, in the end, she just needed to find a way to put their current thinking and behaviors (status quo) at risk. And that she did. When your customers trust you personally and they recognize that you know your stuff professionally, you can begin to communicate in a way that naturally forces them to see things from a different perspective. This helps create urgency to solve problems that they either didn't know they had or knew they had but may be trying to solve ineffectively. This book will be the road map that helps you consistently and intentionally communicate in a way that builds trust and drives urgency on the part of your customers to take action.

The Decline of Traditional Sales Approaches

In today's business environment, the stakes are even higher and the challenges greater than when I wrote the first edition of NeuroSelling. Research from Forrester reveals that 60 percent of B2B buyers now prefer not to interact with sales representatives as their primary information source—up from 53 percent just three years ago.[1] McKinsey's analysis shows that over 70 percent of B2B decision-makers prefer remote or digital interactions, even in a post-pandemic world.[2]

Couple that with the fact that LinkedIn's State of Sales Report shows that salespeople are only spending thirty percent of their time in front of the customer actually selling and you begin to see a major disconnect.[3] Being able to get in front of the customer and having the differentiated, world class communication habits to move them to action is a distinct advantage like never before.

This disconnect between how salespeople traditionally sell and how customers actually make decisions isn't new, but the gap has widened. In a landmark study by Antonio Damasio, subjects with damage to the emotional centers of their brains—rendering them unable to feel emotions but leaving their analytical capabilities intact—were unable to make even simple decisions.[4] Without emotional guidance, the analytical part of the brain becomes paralyzed by options.

Yet most sales training continues to focus on facts, figures, and features—precisely the kind of information that overwhelms rather than persuades. As Daniel Kahneman demonstrated in his Nobel Prize-winning research, all humans have two distinct systems for processing information: a fast, intuitive system that makes most of our decisions (System 1) and a slower, more deliberate system that only engages when absolutely necessary (System 2).[5]

After working in and around biotech for a couple of decades and then partnering with and even hiring world-class PhD researchers who specialize in neuroscience and behavioral psychology, I've had the good fortune and access to scores of data and research on how the human brain works.

You and I might like to believe that in a B2B (Business to Business) setting, we make more reasonable and rational decisions than we would as a typical consumer. The truth is the way our brains work from a decision-making standpoint works the same way, regardless of where we are. We likely won't make an impulse buy of an enterprise software platform, of course, but the way our brains process information still governs how we eventually arrive at our decisions, even when the buying cycle spans months or even years.

The most effective way to impact that decision-making process is to be a world-class communicator.

Why Before What

I can tell you exactly where my understanding of effective communication came from. The farm. And one of my earliest teachers? Papaw Willie Bloomfield.

I'm an old farm boy from north central Ohio. My grandfather—"Papaw"—bought a farm of nearly a hundred acres with his life savings when my dad was just a boy, having moved the family up from Kentucky. With just an eighth-grade education, he was the smartest man I'd ever met. He was an amazing storyteller and communicator.

He taught me how to drive when I was just five years old, standing between his knees on our old, green John Deere tractor. He believed hard work and perseverance will get you everywhere you need to go in life. He believed that problem-solvers rule the world. That with enough curiosity, creativity, and—in our case—maybe a little bit of duct tape, you could solve almost any problem.

He also taught me what I call today the Platinum Rule: *You should treat other people better than they expect to be treated, and it'll always come back to you.*

And whether he was borrowing Old Man Crouse's old red truck from down the road that seemed to frequently be on empty and returning it full of gas or giving his coat off his back, literally, to an older man during a farm sale in the dead of winter, that's just how Papaw lived his life.

He also taught me that family matters more than anything else. That long after our work colleagues and friends are gone, your family remains—so you've got to treat them accordingly along the journey. Papaw came from a long line of storytellers. He could weave a tale like Shakespeare himself, be it about his latest fishing trip or simply to make a point about why we were moving the cattle from one field to the other. People seemed to love listening to him . . . most of all, me.

On a cold February day in 1982, I jumped off the school bus to head down his fifty-yard-long gravel driveway like I did nearly every day before. Instead of seeing just his green Chevy Silverado parked at the end, on this day, it was full of cars. Not too long after, an ambulance came screaming down our snow-covered dirt road and down that same fifty-yard-long driveway.

Unfortunately for me, he had stage 4 lung cancer, and that would be the last day I would ever see him again. I was devastated to lose my mentor, my hero. But what he really taught me was how to take those beliefs and apply them in a way that is meaningful to someone else. To make a difference in the world and in the lives of others. And my Papaw, being a great storyteller, great communicator, and influencer, essentially taught me this methodology long before I was able to validate it with actual science. Thanks in large part to him, I'm able to take these concepts and beliefs and turn them into something that you can hopefully use to make yourself even more effective in the way you communicate . . . be it as a salesperson, a leader, a parent, or the coach of your kid's little league team. And that's what NeuroSelling was born out of. In a lot of ways, I owe a great deal of my company and my life today to my Papaw. And that's really why I do what I do.

With that as the backdrop, what I found anecdotally over the course of my career, as well as pouring over the volumes of published research on the topics of neuroscience and human behavior, is that much of what prevents us from communicating like Papaw may not be your fault.

You've been trained to sell backward.

Neuroscience and Sales: What's in the Mind Is All That Matters

When you were hired, how much of your company's training was devoted to understanding how to build trust faster . . . creating deep and lasting empathetic connections with your customers? I'll bet none. In fact, I've asked a number of CEOs and sales VPs how much training is devoted to interpersonal skills and communication. You know what I hear more often than not? "They're supposed to already know how to do that stuff. That's their job."

Instead, they keep pouring budget into new technology, different sales systems, more data, and failed "sales training du jour" while ignoring the most crucial and yet most overlooked part of a sales/customer relationship: successfully communicating with the customer in a way that builds trust faster and creates urgency to change.

Let me know if this sounds at all familiar:

On your first day, you were probably handed a binder chock-full of product specifications, data, research, facts, and figures along with a binder of policies and procedures (or asked to read the same information online or in your company's boring LMS). You were assigned an area or territory to cover. You probably spent two weeks or more going through online eLearning modules and maybe a couple of months shadowing senior sales reps, followed by a few days of in-person "sales training," and then told to get out there and hustle. We refer to this as the "rep" factory. Every company has a version of this, and, in the end, it tends to spit out the same product (you) with a different label (your brand).

You know the facts, data, features, and benefits cold.

Then, when you want to up your game, you typically turn to sales books or conventional sales "trainers" and inevitably work on refining your sales tactics—i.e., rapport building, body language, mirroring body language, closing "ABCs," social proof, the law of reciprocity, asking probing questions, etc.

Sales tactics? Check. You have those in abundance.

And for a time, many executives thought (and hoped) technology might hold the secret, so they invested heavily in sales software. They found out that tech is a great enabler and potential sales multiplier tool, but it's only as good as its users. Even AI, with all its amazing capabilities and potential, is, in the end, still just a tool. Likely, your company has rolled out programs that infuse marketing into the sales process, leveraging the latest and greatest systems and processes.

Tech? Check.

But let's go back to basics for a moment.

Effective selling really begins and ends with the customer conversation, doesn't it? This is the true "moment of impact."

I hate to be the first person you may hear this from, but what if I told you that your fundamental approach to communicating may be wrong? You were trained to present facts and data, features and benefits—armed with sales tactics and an entire support team behind you. In short, you've done everything you can do to ready yourself to win the customer.

The Trust Deficit in Modern Sales

Unfortunately, customers who reported feeling skepticism in the buying process (i.e., lack of trust) were 164 percent less likely to choose a high-quality, low-regret deal, while customers who felt confident in the information (and sellers) they encountered were 157 percent more likely to purchase.[6] In addition, HubSpot research indicates just 3 percent of buyers trust sales representatives, ranking them lower in trustworthiness than politicians.[7]

This crisis of trust isn't just about perception—it has real consequences. Gartner research shows that B2B customers who receive high-quality information from suppliers are three times more likely to make larger purchases with less regret.[8] Yet most sales presentations fail to provide this information in a way that builds trust or drives urgency.

The paradox is that in an era where information is more abundant than ever, customers still struggle to make confident decisions. This is what psychologist Barry Schwartz called "The Paradox of Choice"—more options and more information actually make decisions harder, not easier.[9]

We will dive into this in much more detail in later chapters, but when we think of the brain, we usually picture the two lobes of gray matter. That's a brain, right? Well, it is, and it isn't. You might find it easier to think of the brain as three brains, all working together to keep you and me alive and well.

Science-Powered Sales Conversations

The two-lobed gray matter you're thinking of is our neocortex, responsible for our higher reasoning. That's where our self-awareness and conscious thought happens. It's what you're using right now to read this book and judge its validity. It's literally how we "think"; therefore, we refer to this area as our "thinking" brain. Another way to "think" about this area of the brain is what Dr. Anthony Jack and many other researchers refer to it, which is the "analytical network." In his research, Dr. Jack found an easier way to understand the brain is to simplify it into two networks, the analytical and the emotional/empathic networks.

The other two parts of the brain that collectively make up the emotional/empathic network hold more sway over our actions and behaviors: the limbic system (made up of areas like the amygdala and hippocampus), which we call the "feeling" brain and what I simply call the root brain, our "instinctive brain" (comprising the brainstem, basal ganglia, and other elements). These two "sub-brains" essentially run in the background, in our subconscious. You don't have to think about making your heart beat. I don't need to think about yanking my hand away from a hot stove. My wife didn't consciously decide to tear up in those moments with the school staff. Our limbic and root brains take care of those types of instinctive and emotional reactions without us needing to consciously think about them.

More importantly for the purpose of this book, **the emotional network (limbic and root brains) initially** *overrides* **our analytical network (neocortex).**

I could be in the middle of an important phone call, but if a car backfires in front of my office, my subconscious immediately overrides my conscious and points my attention to the source of the loud sound. Only after I know that there's no threat can I turn my attention back to my phone conversation. As famed researcher and neurobiologist Joseph LeDoux states in his commentary on the study of emotion:

> *Our studies have led us to understand that the amygdala is the key, no matter how the stimulus comes into the brain: through the eyes, the nose, and the ears. The amygdala is programmed to react without the benefit of input from the thinking part of the brain, the cortex. Eventually, the cortex gets involved, but this processing takes longer.*[10]

You don't need to memorize the parts of the brain to understand the big takeaway here:

Human decision-making *starts* **by processing information in the emotional network (limbic and root brain) and is then** *validated* **by the analytical network (neocortex).**

While we might believe that our decision-making begins in our neocortex (where we become consciously aware of our thoughts), the science you're going to read about in this book shows that it's actually the opposite. Like an iceberg, most of the brain's activity happens below the surface.[11] Our decision-making process starts with the instinctive and emotional filters of our subconscious, then gets passed up to our rational, conscious mind to affirm, validate, or reject that information.

In short, we buy on emotion and justify with logic.

The True Path to Customer Decisions

Recent research from the University of Pennsylvania using functional magnetic resonance imaging (fMRI) has shown that when consumers are presented with brand messaging, the emotional centers of the brain activate before the analytical regions.[12] This sequence is crucial—if the emotional centers don't engage, the analytical areas often fail to fully activate at all.

This mirrors what neuroscientist Antonio Damasio discovered in his groundbreaking research with patients who had damage to their ventromedial prefrontal cortex—the brain region that processes emotional responses. These patients maintained normal intelligence but couldn't make even simple decisions because they lacked emotional input to guide their choices.[13]

A few years back, we had a gentleman in one of our workshops who challenged the idea that we make decisions emotionally and subconsciously and then use our rational mind to validate the emotional decision. He gave me the example of his recent bike purchase. He's an avid cyclist and wanted to upgrade his gear. He combed through all the data on different bikes available, even going so far as to make a spreadsheet with all the specs. He then made his purchase: a pro bike for almost $3,000. He was adamant about the fact that he had made a rational, conscious decision. No emotion there.

"That's great," I said. "So, you're a hard-core cyclist, huh?"

"Aww yeah, man. Every weekend. I bike for miles in some of the most gorgeous places you've ever seen."

"I bet! So, what is it about cycling that you like so much?" I asked.

He got this dreamy look on his face as he began talking about it: "I love it. The freedom, the peace, the road, the connection, and the sense of calm I feel. Seeing the world in a whole different way than you do behind a windshield at sixty or seventy miles an hour."

After he trailed off, I said, "Can I point something out? Everything you've just described is based on emotion. Or, as you called it, irrational emotion. If you simply wanted to exercise, you could buy an exercise bike. No chance of getting hit by a car in your living room. If you simply wanted a bicycle to leisurely ride around and take in the beautiful sites, you can find one at a garage sale for a lot less than $3,000. But you wanted to enhance the emotional experience you get from cycling through the mountains. Cycling isn't about what you do; it's about who you are. It sounds to me like you'd already made an emotional decision to buy a top-of-the-line bike. After that, it was simply a question of which one. You had chosen an expensive bike well before you even began your evaluation. Your 'logical' evaluation was simply an exercise in justifying spending that much money on a bike."

His fellow trainees began laughing as a confused look came across his face. He still wanted to argue that he didn't make emotional decisions, but everyone else in the room saw the truth, plain as day. By the end of the course, he was feeling that way too.

Why do we believe our customers are any different?

The Emotion-Logic Sequence in B2B Decisions

Research from Google and CEB published in the Harvard Business Review found that B2B customers are even more emotionally connected to their vendors and service providers than B2C customers.[14] When asked what drove their business decisions, B2B buyers reported that personal value (such as confidence in their choice or reduced anxiety) had twice the impact of business value.

When you're in a sales conversation, you've been trained to present information to your customer's rational, reasoning, analytical neocortex: facts and figures, features and benefits. You're trying to get them to make a decision by appealing to the most recalcitrant, least persuaded, and least impactful decision-making part of the brain.

That's why we hit a stalemate in the conversation with my son's teachers. My wife and I had been presenting facts, informing them of his condition. Their analytical networks filtered everything we said and silently explained away our concerns, as they already understood all they needed to know about the issue and had no compelling emotional reason to act or behave differently. In other words, there was no problem; therefore, no need for a solution. We could see in their faces and their language that they weren't going to substantially change anything. They weren't going to treat Drew any different than any of the other allergy sufferers who'd passed through the hallowed halls of their prestigious learning institution. They weren't going to treat Hazel and me any differently than any of the other helicopter parents who'd demanded their child be given special treatment.

These were genuinely good people who just weren't connecting with what we were saying. Their neocortex said, "We've already heard all this before. We've had plenty of kids with allergies, and we've always taken care of them. Let's reassure these two parents and get out of here—school opens tomorrow, and we've got a million things left to do!"

What Hazel intuitively did was brilliant. (I'd like to pretend that she's picked up something from hearing me talk about this stuff all the time, but she was that kind of smart way before we ever met.) She used a "neural" pattern interrupter to disengage their neocortex and spike the cortisol, activating the emotional centers of their limbic and root brains. Embedded in that pattern interrupter "story" was an immediate, fear-inducing visual antagonist: a loaded gun endangering a helpless student. Was it an over-the-top analogy? Maybe, but when you are up against the status quo, sometimes you must take drastic visual and emotional measures. She then created a new synaptic connection between the image and the associated emotions of a loaded gun to those seemingly innocent enough peanuts.

Once that new neural pathway was established, the emotional network then overrode the neocortex's attempt at status quo logic. Instead of seeing two overconcerned parents, their neocortex followed the limbic and root brains' direction, which said, "Here are two people trying to save their son from real danger!" It changed everything. It might even have saved Drew's life.

Moment of Impact: Fifteen Minutes of Life or Death

"Jeff, I have to confess: I fully buy into NeuroSelling, but I'm nervous about introducing it at my company. We have in the neighborhood of a million dollars invested in another sales 'system,' and I'm . . . well, I guess I just don't want people to get confused."

I recognized what he really meant. This CEO was nervous about bringing in another sales "program" and running the risk of muddying the waters. Even though they really weren't getting the results they needed, he didn't want to say that he'd potentially made a bad decision with another sales system, especially after investing that kind of money.

"I completely understand," I said, "but NeuroSelling is not a sales system. It's actually a communication methodology. It's about having effective conversations with your customers that build trust faster and drive urgency to change based on the science of human decision-making.

He looked a little confused, which is what I'd hoped for: I'd piqued his attention.

"Let me explain. You sell medical devices, right? What does it look like when your salespeople pitch to a surgeon? Are they sitting in a nice, quiet conference room with a sleek PowerPoint and the surgeon's full attention?"

"I wish!" he laughed.

"Right. Most of the time, they have fifteen minutes—and, really, more like five—between surgeries. The salesperson and the surgeon are standing in a hospital hallway, the doctor is just coming out of one surgery and getting ready to go into another, nurses are running between rooms, lights are flashing, alarms beeping and buzzing, life and death happening all around them, the surgeon's mind is still on their last patient, this is their third sales rep interaction of the day, and your salesperson has just fifteen minutes—and really five—to completely capture their attention."

I gave him a moment to paint that picture in his mind. "If that conversation doesn't go anywhere, if it doesn't captivate the doctor, if it doesn't lead to any action, nothing else matters. Everything in your company hinges on those fifteen minutes. If you want higher sales, if you want to grow your company, if you want to save more lives with your medical devices, then everything you do has to focus on supporting that conversation in that hallway, one-on-one with that doctor. Until that domino falls, nothing else matters. Now, let me ask you this: How do they sell today?"

"Well, they're armed with the data from the clinical trials and—"

"Right there," I stopped him. "Right there: You're talking about facts and figures, features and benefits, why your device is better than what they're using right now. And it is! I've read the data from your clinical trials. It absolutely is. But when you start talking about why your tech is superior to what they've been doing, you've fallen into the trap the majority of businesses fall into—attempting to sell to the wrong part of the brain. I continued, "In fact, they aren't just speaking to the wrong part of the surgeon's brain; they are doing so in a way that causes the surgeon to resist! Essentially, at a subconscious level, they are telling them that what they've been doing is wrong and that they must be some sort of idiot! We human beings like to think of ourselves as pretty good decision-makers. When you layer on a human being who is an educated professional at their position, well, let's just say that person likely feels they know a heck of a lot more than you do. And now you're going to tell me that I've been wrong all this time? Probably not the best way to drive change, huh?"

This CEO didn't need a better sales system; his salespeople just needed to have better problem-solving sales conversations.

Unfortunately, he'd fallen into the same boat as nearly every other company: training on the "hard" stuff (facts and data), assuming too much on the "soft stuff," and having no way to ensure the order in which the conversation took place incentivized change instead of inciting resistance. **But whether you sell medical devices, life insurance, elevators, or envelopes, companies live or die according to what happens in those "moment of impact" customer meetings.**

The High-Stakes Nature of Modern Sales Conversations

The pressure on sales teams has never been higher. A 2023 study by Forrester found that 73 percent of B2B sales leaders reported increasing their revenue targets despite 69 percent expressing challenges with hitting their current goals.[15] At the same time, sales cycles are getting longer and more complex—Gartner research shows the average B2B buying group now involves six to ten decision-makers or stakeholders, each with their own information sources and priorities.[16]

This is precisely why the quality of each customer interaction has become even more critical.

So why don't we typically train our salespeople on how to have incredibly effective conversations?

Every salesperson is trained on coming across with professional credibility. On one hand, that makes sense. It's hard to sell premium engineered machinery, enterprise SaaS, or medical devices if you have no idea what you're talking about, right? But sales organizations almost never train salespeople on how to build trust the right way and in the right order in a way that creates meaningful connection and drives urgency to change.

So regardless of how many VPs of sales tell me, "I hire good, experienced people—they're already supposed to know how to do that stuff. We just train them on our particular product," it doesn't change the fact that this approach simply isn't enough.

Unfortunately, their salespeople came from other companies that assumed the same thing. When were you trained and coached on how to build genuine trust and establish a real human connection? You weren't. And what does your new employer train you on? The exact opposite: facts and figures, features, and benefits. You know how to establish professional credibility on the products and services you offer, but not until someone trusts you personally will their brain allow them to be open to how you may help them professionally. Until you get their stress level down and their feelings of trust and safety up (more on neurochemistry later), he or she won't allow themselves to believe that you're not there simply on your own agenda. Yes, the most important part of the entire sales process—the thing that the entire business hinges on—is what happens in that meeting. No pressure, huh?

Everything in your company, from payroll to R&D to overhead, all rests on what happens in the conversation you and your peers have with the customer and your ability to motivate them to give you money in exchange for your product or service. If that's the most important thing, doesn't it make sense that, above all else, you become an expert at the customer conversation?

It's time to move beyond the traditional, transactional sales rep training "factory" model and move into a new age of communication that is grounded in science and allows you to be authentic, genuine, and still able to drive the conversation in a way that creates urgency to change.

NeuroSelling: A New Communication Approach

NeuroSelling will equip you with a working knowledge of the biology, physiology, and psychology of human decision-making so that you can have more effective and impactful customer conversations. As a professional communicator, you need an understanding of how the human brain works so that you're working with the human mind, not against it.

Let me state what NeuroSelling and the rest of this book *are not*:

- a new spin on old, traditional selling models
- another sales "system"
- just another sales process

This *communication* framework/methodology doesn't need to displace or replace whatever you're doing now. Think of it this way: if your current sales process or sales model were a racecar, NeuroSelling isn't a replacement for that vehicle. It's the rocket fuel to help you win more races. It's a way to shortcut the meetings that lead to nowhere, the months of establishing credibility, the multiple phone calls and meetings that lead to yet more meetings, and sales cycles that last far too long.

I'm going to give you a working knowledge—only what you need to know—about how to lead a sales conversation by speaking to the areas of the brain that really make all the decisions. You're going to experience the powerful advantage of using strategic narratives to engage your customers on an emotional, subconscious level that builds trust and drives change.

The Evolution of Decision Science

Since the first edition of NeuroSelling, the scientific understanding of decision-making has continued to evolve. While the core principles remain valid, new research has deepened our understanding of how decisions are formed in the brain.

Dr. Lisa Feldman Barrett's groundbreaking work on constructed emotion has revealed that emotions aren't simply hardwired reactions but are constructed by our brains based on past experiences, cultural context, and physical sensations.[17] This means that emotional connections in sales conversations must be built with an understanding of the customer's unique frame of reference.

Meanwhile, researchers at Stanford University have used advanced neuroimaging to show that decisions begin forming in subcortical regions of the brain seven to ten seconds before they reach conscious awareness.[18] This research fundamentally challenges the notion that customers consciously evaluate options before deciding.

Wouldn't it be great if you understood exactly what your customer is thinking at all times?

Wouldn't it be fantastic if you knew the process that they go through in their mind in order to make a decision?

What is decision-making, after all?

It's simply the mental process that a person, including your customer, goes through in order to choose between multiple options. Between option A or option B, even if one of those options is status quo—doing nothing different. So how do our customers *choose*?

How do you choose? It's all built-in and hard-wired into our brains. There's a science behind the way every human being makes a decision, yes, including your customers. And when you understand that process, you understand the biology that our brains use and go through to make decisions; what happens as a sales and marketing professional is you can start to craft a more compelling and influential message that hits the part of your customer's brain where they're more receptive and less antagonistic to your message.

We will break the rest of the book into two sections. Part 1: The Lab is all the foundational science and research you need to understand "why" we must communicate differently. Part 2: The Field is your practical roadmap to implement NeuroSelling directly into your customer conversations. Do yourself a favor, overachiever, and don't skip Part 1. Trust me. It's the very information your brain will need to justify moving from your current sales safety box and into a new realm of influence.

I've been fortunate enough to share this NeuroSelling methodology with thousands upon thousands of salespeople all over the world. I've seen how it's changed not only their careers but also their lives. I sincerely hope you experience the same over the course of your journey through this book.

That journey starts next with learning in Part 1: The Lab.

To Learn More About NeuroSelling®, go to
www.braintrustgrowth.com/neuroselling

PART I

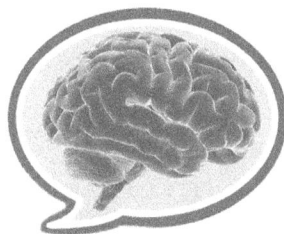

THE LAB

2

SELLING VS. *NEUROSELLING*

"We are not in the coffee business serving people,
but the people business serving coffee."
—Howard Schultz, CEO of Starbucks

THE VICE PRESIDENT of sales for a large B2B software company got up on stage. Behind him, the two massive, thirty-by-fifty screens displayed the tech company's year-to-date and year-over-year sales revenue. In front of him was the company's salesforce of several hundred people. I sat in the very back of the hotel ballroom, waiting for my turn on stage as this year's keynote speaker.

John looked like the stereotypical vice president of sales from central casting—tall, fit, and confident, yet relaxed and charismatic. As he began to give his talk, I could see why he'd been successful and risen to the level he did. He began by showing the usual columns and pie charts representing the sales figures for each of the last three quarters, then showed the gap between that and their annual sales goal. He broke down how much each sales team needed to sell in the fourth quarter to hit the goal. His next few slides of bullet points laid out the types of business clients they were targeting, average closing ratios, and new product specifications and features to use in sales presentations.

All in all, the typical corporate sales talk. Now, I'm not sure if you're aware of this or not, but a roomful of salespeople is a pretty hard audience to keep engaged. You've heard the old phrase, "The mind can only absorb what the butt can endure," right? The longer John talked, the more people began fidgeting in their seats, checking their phones, and whispering to their coworkers beside them. Toward the back, where John couldn't see them, I even saw some on Facebook and playing Candy Crush. He wrapped up his talk with the typical Knute Rockne rallying cry of "Okay, let's get out there and crush Q4!" Everyone provided the customary polite golf clap. (Of course, they did; wouldn't you applaud your boss's boss's boss?) Then John introduced the next speaker, scheduled to go on just before me. "Alright, now Christina Yearwood" (not her real name, so don't go googling her) from corporate HR is coming up to talk about our leadership development program."

The thirty-something woman who bounded up on stage was as excited as she was nervous. She thanked the VP but then fumbled a little in her lead-in. After a minute, she began to find her rhythm, and instead of describing the details of the leadership development program she was there to "pitch," she instead singled out a couple of people in the room who'd already signed up and then mentioned it was the single most impactful program she had ever personally been through in her entire career. It sounded like a solid program, and I saw more than one head in the room nodding along. Then Christina played a video of testimonials from other people around the company who'd gone through the training. It seemed like I'd sat through a million of these, so I thought I knew what to expect, but I was pleasantly surprised. Instead of the usual generic praise, the people on screen were relating some personal and emotional experiences they'd had while going through this particular leadership development program and how it had impacted them as much on a personal discovery level as it did professionally. I thought, *That's pretty cool*. Most companies don't go there. They like to keep things strictly business and don't encourage sharing too much personal information. The fact that this company had a leadership development course touting these types of experiences and benefits said a lot about how much they cared about their people as individuals.

Usually, this kind of pitch would end there. The person from corporate would wrap it up with directions and details on how to sign up and then thank the audience for their time. After that, it would be my turn. I had already half stood up from my seat when Christina said, "My own personal journey through this program was one of amazing self-discovery, and I'd like to share my story with you." I thought, *Okay, this is different*. I sat back down in my chair and waited as she gathered her thoughts. What happened in the next few minutes transformed what should have been a simple training pitch into an experience I doubt any of us in that ballroom will ever forget.

An Unscripted Experience of NeuroSelling

"When I was a child, I was the middle daughter of three girls. My father, however, had always wanted a boy. It seemed like every day, he somehow found a way to remind us that we weren't what he wanted. He and my mother had tried three times and been disappointed every time. 'Three strikes and you're out,' he used to say to us. If we had been born boys, he'd say, he could take us to a ball game or work on cars together. But since we were girls, we weren't good for anything. The rest of the time, he just ignored us," Christina began. "As the middle child, I felt especially neglected. I've learned now that experts call it 'middle child syndrome.' My oldest sister got most of my mother's attention because Mom depended on her so much; she was 'Mommy's little helper.' My youngest sister got attention because she was 'the baby,' and we needed to 'watch out for her.' I just kind of got stuck in the middle." Christina had to fight back some tears before continuing on. "All my life, I felt like I was never good enough. I was so heartsick over being born a girl. 'If only I'd been born a boy,' I used to tell myself, 'my daddy would love me.' But it didn't matter how hard I tried or what I did; it was never enough. I grew up being ashamed for just being alive—for just being who I was.

"When I was a freshman in college, my dad walked out on us. No calls, no emails, no texts. He just left. We were all devastated. I'd never really been close to my family anyway, but instead of pulling us together, Dad leaving somehow pushed us all further apart. There I was, hundreds of miles away from my family, my father had just walked out of our lives, and my mother and sisters became even more distant. I had never been so lonely and heartbroken in my entire life."

I had become so caught up in her story that I'd completely forgotten everyone else sitting there in the ballroom. I came back to earth for a moment and saw everyone in the room entranced, leaning forward in their seats, hanging on her every word. We could all picture Christina in her college dorm crying. We could feel her heartache. Intellectually, I recognized what was happening. Christina's story had bypassed our neocortex, responsible for our higher-level thinking, rationalizing, and judgment. Instead, she had engaged our limbic systems and root brains, responsible for personal trust, empathy, personal relationships, and a lab's worth of neurochemicals like dopamine and oxytocin. Despite knowing the neuroscience behind what was happening, despite teaching people to work *with* how our brain is designed (instead of against it), despite having an entire business dedicated to neuroscience selling techniques, despite the professional in me recognizing exactly what was going on—there I sat, in the back of the room, eyes filling with tears along with Christina's and engrossed in her story like everyone else! But Christina quickly drew me back into her narrative.

She had paused for a moment, but then her entire body language changed as if she were preparing for something. Her voice lowered as she dropped the mother of all bombs. "That loneliness and heartbreak . . . I couldn't take it. I tried to kill myself." The temperature of the entire room dropped several degrees. I heard people throughout the audience react, some sucking in their breath or letting slip an involuntary "Oh, my." Christina said, "The only reason I'm alive is because my roommate found me and rescued me in time. After getting out of the hospital, I went into recovery for suicide victims. Through my experience, I met a man passionate about suicide prevention, the man who eventually became my husband. He's passionate about it because he lost his own roommate to suicide in college. With his love and understanding, I went through therapy, fully recovered, and reentered the workforce seven years ago. But, guys, let me tell you what this leadership development program did: It made me look inside myself. It held a mirror up to who I am, not only as a professional but also as a person. And you know what? It helped me realize how I'd always unconsciously seen myself—as unworthy. This program helped me find a vision of who I'd like to be. Because of those two realizations, I was finally able to let go of all that hurt. I let go of those feelings of inadequacy. It affirmed my self-worth. I found out that I didn't need to feel like I wasn't good enough or that I'd never fit in. It gave me the sense of self I'd always been missing without even knowing it. I discovered that I can write the rest of my life story however I want to."

She smiled a beautiful, dazzling smile and said, "And I found a new purpose in life. I want to help other people realize that they don't have to live with the identity they've carried all their lives and that they can define their own self-worth and purpose. That everyone's been created in a unique way and for a unique purpose. They need to find it and walk in that purpose." With a tear-streaked face, she closed with, "That's what this program did for me. I would love it if you'd let it do something similar for you too. If you'd like to learn more about the details of the program, here's a link, or you can see me at the break. Thank you."

The room exploded. People were on their feet, cheering. The applause was deafening. Many were freely crying. I have never seen a standing ovation before or since for a leadership program commercial. What an unforgettable moment. John took back the mic and said, "Wow! What a story. And now for our speaker!" I mean, come on! How do you follow something like that!?

The Science Behind the Perfect Pitch

Fortunately for me, I could not have had a better setup. It was like I had paid Christina to work her magic in preparation for what I was about to share. Unbeknownst to them, my entire talk revolved around creating exactly that kind of magic in client sales conversations. For the next hour, we talked about how to use neuroscience to do intentionally what Christina had done intuitively (and, most likely, accidentally). After the talk, I was having lunch in the adjoining ballroom with a group of those salespeople. Of course, they were talking about how moving Christina's story was and how intriguing it was to have a presentation straight afterward explaining the science behind the experience.

In between forkfuls of macaroni, I said to the whole table, "Let me ask you something: Do you remember the bullet points on slide number four of John's presentation?" They all stopped for a minute while they racked their brains. Someone said, "I think it was something around how much more everybody's team has to sell. Right?" "Okay," I said, "what else? Remember any of the other bullet points? Any of the numbers from his graphs and charts?" After everyone kind of shrugged, I said, "Okay, how about the main takeaway?" "He just, I guess . . . he just wants everyone to sell more? That we need to bridge the gap in Q4?" one ventured. "What's this pop quiz for?" "Pause for a second. Let me ask you this: What do you remember of Christina's story?" Everyone started chiming in all at once: her suicide attempt, her husband's roommate, she had two sisters, her mother doting on the baby, her roommate finding her in her dorm room, her awakening and walking in her purpose.

All of them had near-perfect recall of her entire story. "Now, what was meaningful from John's talk?" I asked. Same answer as before: "That we need to sell more." I said, "So two people pitched two business ideas to you guys this morning that required you to potentially 'change' or do something 'different.' One person was an experienced sales professional with years of sales training under his belt. He was selling you on the idea that you need to keep doing what you've been doing—just do it better. Right? "The other professional," I continued, "doesn't have a sales background. In fact, she's from HR. She got up to sell you on the idea that you needed to do something completely new and different—something that would make you uncomfortable, force you to change, and challenge your beliefs about yourself. Now, contrast those two sales pitches. Which one were you actually sold on?" I mean, there wasn't a question. It wasn't even a contest. Christina's. "Last question," I said. "Think about yourself in sales conversations in front of your customers. Which talk looks like what you do? John's? Or Christina's?" Again, no question: definitely John's. Smiling, I said, "You see the problem, right? You were convinced to change by Christina, but you sell like John!"

The Neuropsychology of Persuasion

What happened in that ballroom illustrates what neuropsychologists have been documenting for years. Christina's personal narrative activated what neuroscientists call "neural coupling"—a phenomenon where the brain activity patterns of listeners begin to synchronize with those of the speaker. Dr. Uri Hasson of Princeton University found that during effective storytelling, the same brain regions activate in both the storyteller and the listener, creating a remarkable mind-to-mind connection.[1] This neural coupling doesn't happen with bulleted lists or data presentations. It specifically occurs when we share relatable narratives with emotional content. When Christina shared her story of childhood rejection and ultimate redemption, she wasn't just transmitting information—she was creating a shared experience that physically changed the brain states of everyone in the room.

Research from the Center for Neuroeconomics Studies at Claremont Graduate University has shown that emotional narratives trigger the release of oxytocin, often called the "trust hormone," which increases empathy and connection.[2] Dr. Paul Zak, the Center's director, found that character-driven stories with emotional content consistently cause oxytocin synthesis, directly affecting how willing we are to trust the speaker and take action on their recommendations. Meanwhile, John's presentation primarily engaged the analytical centers of the brain—regions associated with logical processing but not emotional engagement or memory formation. Dr. Antonio Damasio's research has repeatedly shown that without emotional engagement, information rarely converts to long-term memory or motivates behavioral change.[3]

The Neuroscience of Memorable Communication

The contrast between John and Christina's presentations also illustrates why certain communications stick with us while others fade almost immediately. Dr. Carmen Simon, a cognitive neuroscientist who studies how the brain processes business communications, has found that people typically forget 90 percent of what they communicate within a week.[4] Her research reveals that what makes the remaining 10 percent memorable is often its emotional impact, not its logical coherence. This memory effect is directly tied to how different types of information engage distinct brain systems. Facts and figures primarily activate the analytical network and dorsolateral regions associated with working memory—areas that have limited capacity and retention. Emotional narratives, by contrast, engage the emotional network, specifically the hippocampus, amygdala, and related limbic structures that are central to long-term memory formation.

In a notable 2011 study published in Perspectives on Psychological Science, researchers Mather and Sutherland demonstrated that emotional engagement significantly increased memory retention of key information, with emotionally arousing content recalled up to 40 percent more effectively than neutral, rationally presented material.[5]

The implications for sales are enormous: if your customer can't remember what you said a week later, how can they possibly advocate for your solution in internal meetings when they attempt to persuade their peers? The "Christina Effect," as we might call it, works because stories create what neuroscientists term "episodic memories"—recollections that include not just what happened but how it felt to experience it. When Christina described her childhood rejection, her college dorm room, and her moment of transformation, each listener created their own mental movie, complete with emotions, visuals, and sensory details. These rich, multi-dimensional memories are far more persistent than semantic memories (simple facts) like the quarterly sales targets John presented.

Conducting Your Sales Conversations in the Right Order

It's not their fault, though. Corporate sales training mostly consists of equipping the salesperson with facts and figures about the product you are selling. When I was selling life-saving oncology drugs, I was taught everything from drug specifications to cell biology, cell proliferation, randomized, placebo-controlled phase, three study data and on and on and on. It took time, trial and error, and a lot of research before I realized people need more than that. Yes, we need the facts and figures from John's talk. If you walk into a prospect's office and do nothing but appeal to them on an emotional level, they might like you, but you'll have zero professional credibility. On the other hand, if you go into that same office and do nothing but recite facts, totally ignoring the emotional human factor, then you haven't established any authentic connection.

The Problem with Traditional Sales Approaches

The fundamental problem with traditional sales approaches is that they're built on a misunderstanding of how human beings process information and make decisions. The standard sales model typically follows a structure like this:

- Step 1: Build rapport
- Step 2: Needs analysis
- Step 3: Present solution
- Step 4: Handle objections
- Step 5: Close the sale

This approach assumes that buying decisions progress linearly and logically. However, neuroscience research overwhelmingly contradicts this assumption. Studies using fMRI scanning technology reveal that decision-making is far from a linear, logical process. In fact, Baba Shiv, professor of marketing at Stanford Graduate School of Business, found that when the emotional centers of the brain are damaged, seemingly simple decisions become impossible to make.[6] His research demonstrated how participants with intact emotional processing could quickly choose between two similar products, while those with damage to emotional centers became paralyzed by analysis. This explains why traditional "feature dumping" approaches to sales are so ineffective. When salespeople present lengthy lists of product specifications and benefits without first establishing an emotional connection, they're actually making it harder for customers to decide.

The Sequencing Effect in Persuasion

Interestingly, the order in which you present information also fundamentally changes how it's processed—a phenomenon psychologists call the "primacy effect." Information presented first creates a mental framework through which subsequent information is filtered and interpreted. When it comes to making information stick, the order of presentation can make all the difference.

A compelling 2003 study by psychologists Elizabeth Kensinger and Suzanne Corkin, published in Memory & Cognition, offers a striking example of this effect.[7] In their experiment, participants were shown a sequence of images—some emotionally charged, like a menacing snake, others neutral, like a plain chair. The twist? When the emotional images came first, they didn't just grab attention—they turbocharged memory for what followed. The researchers found that emotionally evocative content was recalled up to 30 percent better than neutral content after a delay, and here's the kicker: this emotional jolt also boosted memory for the neutral details that came next. It's as if the initial wave of feeling primed the brain to lock in the drier, more analytical facts that followed.

Imagine a presenter starting with a heartfelt story before diving into technical specs—the audience doesn't just feel more; they remember more. This isn't about overwhelming with sentiment; it's about setting the stage so the mind can process and retain the complex stuff with greater clarity. Kensinger and Corkin's work shows us that leading with emotion isn't just persuasive—it's a memory amplifier, making the technical bits far more unforgettable.

This sequencing phenomenon explains why Christina's approach works so well. By using her personal narrative, she created what psychologists call a "cognitive scaffolding"—an emotional framework that helps listeners organize and make sense of any subsequent information. Had she begun with technical details about the program and only shared her story at the end, the impact would have been dramatically reduced. Research by behavioral economist Dan Ariely offers another insight into this sequencing effect. In what he called the "predictable irrationality" of decision-making, Ariely found that initial emotional impressions create an "anchor point" that significantly influences all subsequent judgments—even highly technical ones.[8] When Christina created a powerful emotional anchor through her personal story, she established a frame through which everyone in the room interpreted everything else about the leadership program.

Why The Order of Information Matters

A critical finding from neuroscience research is that the sequence in which information is presented fundamentally changes how it's processed. Dr. Robert Cialdini, in his landmark work on persuasion, demonstrated that the order of information dramatically affects its impact.[9] When facts and figures are presented first, the brain's skeptical analytical network activates, creating resistance. But when the same information follows an emotional connection, it's processed through different neural pathways and encounters significantly less resistance. This sequencing effect is why Christina's emotional story was so much more impactful than John's data-driven presentation. Her personal narrative activated the brain's emotional and empathic networks first, creating an open and receptive state for any subsequent information.

The Sales Trust Matrix

This creates what I call the Sales Trust Matrix—a framework for understanding how customers perceive salespeople based on two dimensions: personal connection and professional credibility. The image in the bottom-left quadrant is the low-connection, low-credibility salesperson—the "incompetent amateur," someone who neither connects personally nor demonstrates expertise. These salespeople struggle to gain any traction with customers. The bottom-right quadrant represents high personal connection but low professional credibility—the "likable amateur." Customers enjoy talking with this person but don't see them as capable of solving complex problems. The top-left quadrant shows the low-connection, high-credibility salesperson—the "competent jerk." They know their stuff, but customers don't connect with them personally. The top-right quadrant—high connection and high credibility, "the credible connector"—is where NeuroSelling aims to position you. This is the ideal position where customers not only like you personally (true connection) but also respect you professionally (perceived credibility).

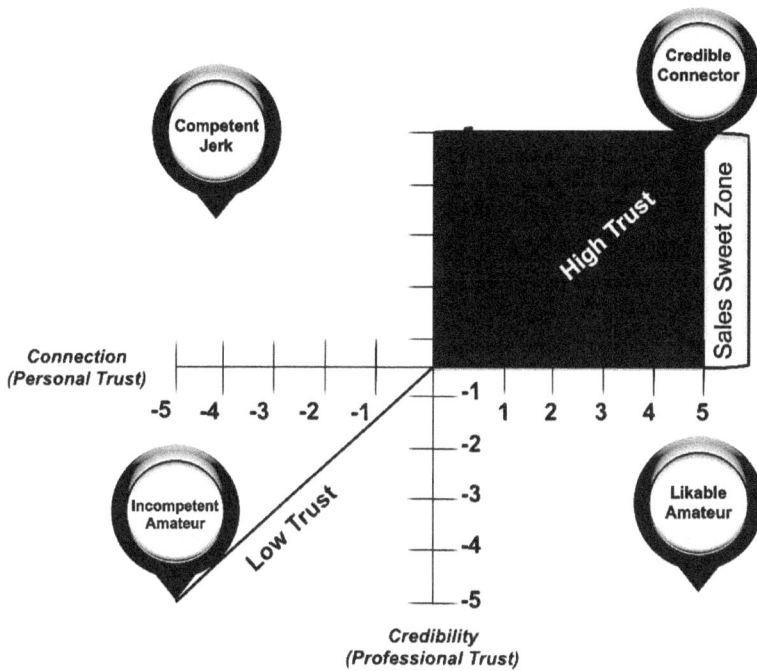

Research by social psychologist Amy Cuddy at Harvard Business School confirms this framework. Her studies show that we evaluate others primarily on two dimensions: warmth (connection) and competence (credibility).[10] Critically, her findings reveal that warmth judgments always precede competence judgments—meaning we decide if we like someone before we decide if they're competent. Cuddy's research has been extended by organizational psychologists studying business relationships specifically. A Kellogg School of Management study of B2B relationships found that customers who ranked vendors high on both warmth and competence were 63 percent more likely to share confidential information and 38 percent more likely to advocate for them internally.[11] These are precisely the behaviors needed to close complex B2B sales. This research lends scientific validity to what may seem like common sense: people do business with people they both like and respect. But the critical insight is the sequencing—connection comes first and fundamentally shapes how competence (credibility) is perceived.

But the truth is—the "awesome" facts and figures that you recite to your customers don't make the sale because they don't create trust on a deeper level. So, *what does*? Sales needs an approach that aligns how we communicate with how the brain *actually* works. An approach that establishes personal trust (selling to the emotional network) as well as professional credibility (selling to the analytical network). But for at least two thousand years, Western culture has had it backward: We try to persuade people by starting with facts and figures. **We've been selling from the outside in when we need to sell from the inside out!**

Back to my lunchtime experience with John and Christina's team. After I finished eating, I pushed my plate back and said, "Let's think about it this way. What if Christina had gone first, connected emotionally to all of us like she did, talking about how you can make a bigger difference through how you're uniquely wired to solve problems for your customers. *Then* John got up and said, 'Here's how we can solve those problems for our customers, here's the gap, here's who you need to talk to, and here's what to say to them.'" I let them paint that picture in their mind, and then I said, "Same graphs, same tables, same information. How do you think it would have changed how you experienced John's talk?" "Man! It'd be the difference between night and day!" a woman at the table said. "After Christina was finished, I was ready to go out and conquer the world!" Exactly.

By the time you're finished reading this book, I want you to understand that the traditional sales training you've had teaches you to engage with only one part of the brain. Unfortunately for you, that so happens to be the hardest, most resistant, most skeptical, and least willing to change part of the human mind. To sell more and more easily, you need to engage with the other two parts responsible for emotion, empathy, visualization, trust, and connection. You also need to learn to sell in the right order: from the inside out.

To understand the human brain, let's go back to how it all began.

3

THE CAVE-TO-CAVE SALESMAN

"People will teach you how to sell to them if you'll pay attention to the messages they send you."

—Jim Cathcart, Relationship Selling

IN MANY WAYS, one of the simplest ways to understand the buying brain is to watch the movie *The Croods*. It's a story about a caveman family, led by the father, Grug Crood. He protects his family by adhering to a long list of rules, the most important of which was this: Anything new is bad.

New animal? Bad.

New person? Bad

New idea? Very bad.

Of course, there's a cataclysmic event and the family is forced to flee everything they've known and embrace change as a way of life. In the end, Caveman Crood realizes that change doesn't always have to be frightening.

That status quo can be limiting to the very goals you have in life and that embracing change can not only be more effective, but it also can even be fun.

It's been a few thousand years since we lived in caves, but our brains haven't quite caught up yet. Millions of years ago, our higher-level "thinking" brain, our neocortex, may not have been as finely tuned as it is today. After all, there were far fewer websites to peruse or new iPhones to master. However, our emotional and instinctive brains, our limbic system and root brain areas were essentially the same as they are today. Those parts of our brain keep us alive. Not only do they control breathing, blinking, and other involuntary things our bodies need, but they also control our instinctive and emotional reactions.

Nobel-prize winner Daniel Kahneman in his book *Thinking, Fast and Slow*, as well as Malcolm Gladwell in *Blink* do a fantastic job of showing how much of our thinking happens in the subconscious—including judgment decisions. For instance, have you ever had a gut feeling about someone? You couldn't put your finger on it, but something felt—off? On the other hand, have you ever felt an instant connection with someone? Played a hunch? Fallen in love?

Those aren't rational decisions. If pressed, you could probably come up with some reason or excuse why you felt the way you did. But that approach ("Here's a decision—now, let me figure out how to justify it.") runs contrary to how we're "supposed" to think.

Understanding Decision-Making Through Brain Structure

Our brains operate in a business environment vastly different from the natural contexts in which humans developed. For most of human history, people lived in small tribes as hunter-gatherers, facing threats from predators, rival groups, and harsh natural conditions. This historical context shaped cognitive patterns that continue to influence modern decision-making.

In the 1960s, neuroscientist Paul MacLean's triune brain model provided a framework for understanding brain structure and function.[1]

MacLean proposed that the human brain consists of three primary functional areas:

1. The reptilian brain or "R-complex" (brainstem, cerebellum): According to MacLean, this area controls basic survival functions like breathing, heart rate, and instinctive reactions.
2. The paleomammalian brain (limbic system): MacLean believed this region governs emotions, motivation, and social behaviors.
3. The neomammalian brain (neocortex): In MacLean's model, this area enables higher cognitive functions like abstract thinking, planning, and language.

The Outer Layer—the funky, folded, gray layer that we sometimes see on TV and in popular movies—is the neocortex.

The challenge with the neocortex is that it also tends to be the skeptical and judgmental part of the brain. As it receives that information, it judges it immediately. Data is either verified and validated or invalidated and dismissed.

The Second Layer—When we go down one layer from the neocortex, we see the limbic system/brain.

If the neocortex is your "thinking brain"—or perhaps another nickname for sales would be the "skeptical brain"—the limbic system is your "feeling brain."

This is the area my wife activated with her "peanuts = loaded gun" analogy. She triggered an emotional response to information the educators already had. But with a new emotion associated with it, their brain processed the exact same information differently!

One could argue that everything that's ever happened to you since you were in the womb gets encoded on little "neuron microchips" and stored in your limbic system. That information helps you or hinders you in how you perceive what's happening in your world today. So, believe it or not, things that happened to you all the way back when you were three or four years old got encoded—stuffed into what I call the "junk in your brain trunk." As you perceive information in the world around you today, information even that far back can and does guide your day-to-day emotional reactions and decision-making process.

The Third Layer—When you go down a layer deeper still, you've got the "root brain," which we call the "instinctive brain": the cerebellum, brain stem, and other faculties in your mind where all your unconscious behaviors spring from. Things like breathing, hunger, avoidance, thirst, survival—all of those mechanisms and responses are being delivered through your root brain mechanism. Some researchers refer to this as the "R-complex" or "reptilian" brain.

While contemporary neuroscientists recognize more nuanced brain organization than MacLean's model suggests, his perspective helps illustrate why our decision-making often seems at odds with purely "rational" processes.

When a snake appears on a hiking trail, we jump back before conscious thought registers the danger. This "affect heuristic"—allowing emotions to guide quick judgments—saved our ancestors from predators but can create problems in modern contexts like evaluating sales proposals.

School teaches us to gather information in order to make an informed decision. We're supposed to "reason it out." Western thought and philosophy rest on this idea, reflected in the famous quote, "I think, therefore I am." This line of reasoning by René Descartes harkens back to the Greek philosopher Aristotle. These two men and those who followed them saw the mind and body as two distinct entities. Consciousness (or rational thought) existed on one level; our bodily needs and functions existed on a lower plane of existence.

Until just a few years ago, many people, including notable researchers, thought that humans used logic to make rational decisions. They, like many others, believed we might lose our temper or become infatuated and make an emotional decision, but once we were clear-headed, we'd return back to our baseline: reason.

With this outdated view, the way we've traditionally been taught to engage in sales and marketing makes sense. We present facts and figures, using logic and reason to convince a customer that our product is the superior choice. In other words, we appeal to their conscious mind, yet they often don't make the logical decision we thought they would, do they?

The West has been wrong for over two thousand years. It's only been in the last few decades that we've come to understand just how inextricably linked our minds and bodies are. What scientists and researchers now understand is that what we think of as consciousness happens primarily in just one part of the brain.

But most of the activity in our brain actually takes place in what we know as the subconscious.

We don't make rational decisions free from emotion and instinct.

Especially in a sales setting, we'd like to think that we're pretty smart, like a modern René Descartes. The truth is we're more like our caveman brother, Grug Crood.

The Dual Processing Theory of Decision-Making

While MacLean's triune brain model provides a useful conceptual framework, contemporary neuroscience has evolved to what's known as dual-process theory. This model, championed by Nobel Prize winner Daniel Kahneman, describes two systems of thinking: System 1 (fast, intuitive, emotional) and System 2 (slow, deliberative, logical).[2]

System 1 operates automatically with little effort, generating impressions, feelings, and inclinations. System 2 requires attention and effort, allocating mental resources to complex computations and conscious choices. The critical insight is that System 1 is always running in the background, while System 2 activates only when necessary—and typically to rationalize decisions that System 1 has already made.

This creates what psychologists call "choice-supportive bias," wherein we unconsciously distort our memories of chosen and rejected options to make our past choices seem better than they actually were.[3] Once your customer has chosen a course of action—even maintaining the status quo—they will unconsciously filter information to support that choice.

Recent research using electroencephalography (EEG) has shown that this bias is not just post-decision rationalization; it affects how information is processed at the earliest stages of perception.[4] This means that by the time your sales presentation reaches your prospect's conscious awareness, it has already been filtered through layers of unconscious biases.

Say I just gave you a big equation: I want you to take 757,625 and divide the answer by 3. Well, sure, you could do it. To do so, your neocortex would be firing on all cylinders, essentially shutting off your limbic system and root brain because you're doing a mathematical equation.

Now, if I told you a factual data point such as, "At Braintrust, our clients see 150 percent increase in sales year over year. We have a 97 percent customer retention," In that moment, your neocortex processes that information, and since it's data/facts, it's deciding rather quickly whether it believes it or not. If you know, like, and trust me, your brain will use those numbers as support for your belief (cognitive bias that we'll get into later), but if you don't know, like, or trust me yet, your brain will likely use those very same numbers as a reason to distrust or potentially disbelieve me.

It takes a lot more horsepower or energy in your brain to start with the analytical network (neocortex) and have it process this vast amount of facts, data, and information. If I get you emotionally engaged and tell you why I started Braintrust, and then tie it to why that matters to you followed by a compelling narrative around your challenges, something changes.

When you are following that type of emotionally connected narrative, you suspend judgment of the facts.

It's important for us to understand, as professional communicators, how the message we deliver is being received and perceived by the brain of our audience.

If you're the type of salesperson who communicates primarily with facts, data, and figures, you're pummeling your customer's analytical network (neocortex).

The Two Brain "Networks" Impact on Decision-Making

One of the most significant discoveries in neuroscience over the past two decades has been the identification of two large-scale brain networks that operate in opposition to each other: the Default Mode Network (DMN, also known as the "emotional/empathic network") and the Task-Positive Network (TPN, also known as the "analytical network").[5]

The Default Mode Network (emotional/empathic network) activates during introspection, empathy, social thinking, and creativity. It's associated with emotional processing, narrative comprehension, and relationship building. In contrast, the Task-Positive Network (analytical network) activates during analytical problem-solving, logical reasoning, and focused attention to external tasks.[6]

Research by Dr. Anthony Jack at Case Western Reserve University has shown that these networks suppress each other—when one activates, the other deactivates.[7] This finding has profound implications for sales conversations. When you bombard your customer with analytical information, you're activating their TPN (analytical network), which suppresses the DMN (emotional/empathic network)—the very network needed for trust-building, emotional connection, and creative problem-solving.

Jack's research helps explain why traditional feature-benefit selling fails to create a lasting impact. By activating the customer's analytical task network first, you've made it neurologically more difficult for them to access the social-emotional network needed for trust and connection.

Information processing begins in the root brain. And it begins with the most basic question: *friend or foe*? Then the root brain passes its determination along to the limbic brain, which makes a determination: *fight or flight*? Not until it clears those gatekeepers and gets the green light does the information make its way to the neocortex for validation or further review.

To summarize, we began with the triune brain approach (Neocortex, Limbic system, R-complex/Reptilian brain) then ventured into Kahneman's "System 1 vs. System 2" and then landed on Dr. Jack's "Analytical vs. Emotional/Empathic" networks. To simplify:

- analytical network (neocortex) = your " thinking brain" (facts and data, skepticism, judgment, evaluation)
- emotional/empathic network (limbic = your "feeling brain" (emotion, memory, internal visualization) and root = your "instinctive brain" (safety, hunger/thirst, avoidance, survival)

When you go into a sales conversation, you need to think of these in reverse order: **Engage with the emotional/empathic network, then—and only then—pitch to the skeptical analytical network.**

When we begin with this "inside-out" approach, we are certainly communicating in the right order, but there are a few more decision-making landmines we need to be aware of.

The first is to understand the "self-preservation orientation."

Self-Preservation Orientation

Have you ever accidentally run into a good friend at the store or one of your favorite customers at the airport? You know when you recognize them and you automatically get a smile on your face because you're genuinely happy to see them? You smile because you feel happy; you're happy because your brain is flooding you with positive trust chemicals and your brain triggers those chemicals because it recognizes a friend.

The Neuroscience of Friend vs. Foe Detection

This automatic friend-foe response operates at a neural level through what scientists call the "threat appraisal system." Research using functional magnetic resonance imaging (fMRI) has revealed that when we encounter strangers, our amygdala—a key structure in our threat detection system—shows heightened activity compared to when we recognize friends.[8]

This threat response happens in milliseconds, well before conscious awareness. Studies at New York University showed that the brain can categorize a face as threatening in as little as 33 milliseconds—faster than you can blink.[9] This rapid categorization triggers a cascade of neurochemical responses that prepare the body for potential danger.

When the brain identifies someone as a "friend," it releases a cocktail of bonding neurochemicals, including oxytocin, serotonin, and dopamine. These chemicals create feelings of trust, pleasure, and connection. When it categorizes someone as a potential "foe," it releases stress hormones like cortisol and adrenaline, putting the body in a defensive posture.

Every moment of your day, your caveman brain is continuously scanning your surroundings, constantly asking: "Friend or foe? Friend or foe?" By nature, everyone is either "foe" or, at best, "neutral leaning toward foe" until our brain moves them into the "friend" category. In other words, potentially "unsafe" until I can determine you as "safe."

When you walk into a sales meeting, your customer's caveman brain already sees you as a potential threat.

The fact that you're there to sell them something only increases the threat level. You're the bad guy, there to get their money. You're already starting off at a disadvantage: As Captain Kirk used to command when the Starship Enterprise was in danger on *Star Trek*, "Shields up, Mr. Sulu! Red alert!"

When the brain recognizes a potential threat, it triggers your adrenaline and spikes your cortisol (stress levels), which is your brain's way of ensuring blood flows to your muscles to allow you to either stay and fight the threat or flee from it.

As the caveman brain goes into fight-or-flight mode, it redirects focus away from the reasoning neocortex to the fear-driven limbic system and root brain. Knowing that, doesn't it make sense why so many sales meetings feel tense or even confrontational? According to the caveman brain, it is! As humans, our instincts in these situations are to amplify "self-preservation orientation." Shields up and phasers at the ready.

Now, "Captain Kirk" might listen to your sales spiel and hold off on firing phasers—he might even ask Mr. Spock about the logical course of action—but his shields are still up. Even if your customer is sitting in a Silicon Valley startup office that looks cooler than the Enterprise, his brain still acts like it's in the savanna, worrying whether a saber-toothed tiger is about to eat him.

But realize this: You do the same thing!

Your Neurochemistry in Sales Meetings

Sales meetings are stressful.

Even if you've been selling for so long that it doesn't really seem like stress, you can't fool your caveman brain. It knows exactly what this is. This is you hunting woolly mammoths, trying to survive another winter. The customer has money you need. It doesn't matter that we've gone from caves to houses and from spears to emails. Your caveman brain still automatically goes into flight-or-fight mode.

The Neurobiology of Sales Stress

When the human brain is under stress—regardless of which side of the desk we're on—our neurochemistry changes. Our bodies are preparing us to either fight the threat or run away to live another day. Just like for your customer, your higher brain functions in the analytical neocortex are impaired as the brain shifts focus to the limbic system and root brain.

This biological stress response creates several challenges for salespeople:

- Executive Function Impairment: Under stress, the prefrontal cortex—responsible for strategic thinking and verbal fluency—shows reduced functionality.[10] This explains why even experienced salespeople sometimes fumble their words or forget key points during high-pressure moments.
- Memory Formation Disruption: Moderate to high stress levels interfere with the hippocampus, impairing your ability to form new memories and recall specific information.[11] This is why you might forget important details about your product or the customer's needs during tense negotiations.
- Empathy Reduction: Stress hormones reduce activity in brain regions associated with empathy and perspective-taking.[12] This makes it harder to read the customer's emotional cues—precisely when you need these skills most.
- Risk-Aversion Increases: Under stress, the brain becomes more sensitive to potential losses than gains, leading to more conservative decision-making.[13] This can cause salespeople to retreat to "safer" talking points rather than addressing challenging topics that might actually advance the sale.

Once we leave the sales meeting, we've all had those "Aww, man, why did I say that?!" moments. When our caveman brain perceives that the threat is gone and it can calm down, our higher-level thinking brain can switch back on. We can reason again and think through what we said and did during the conversation. But in the moment, we panic, and when we panic, we revert back to what we know.

One of my sayings is that **in a stressful situation, people revert to communicating from their habitual highest level of training, knowledge and experience**. This is why the military trains our soldiers the way they do. In the heat of battle, soldiers can't rely on their neocortex. They have to rely on their ingrained training to see them through. On naval ships and submarines, they're constantly running practice drills. If a real fire breaks out on a boat, each sailor's fight-or-flight instincts kick in. The only way they can function flawlessly is by falling back on their instincts and muscle memory.

And what does the typical sales organization train or "ingrain" into their people?

Facts and data, features and benefits.

Whenever those same people go into a meeting that their income and livelihood depend on, their stress hormone cortisol spikes, and they revert to their ingrained training.

No wonder things never change—self-preservation is a basic human instinct.

From a survival of the species standpoint, it's certainly the most important. When we experience pain or fear, we revert to our basic instincts: How can I survive this? What's safe? Our walls go up, and our defense mechanisms kick in. Our instincts are to stick with what's safe and comfortable—what we know. Fortunately for us today, what we "know" may not get us killed like in the savannah, but unfortunately for many of us, it leads to the death of our sale, nonetheless.

The False Security of Your Safety Box

On behalf of a medical device client, I once attended a medical conference with a group of surgeons in Denver. Over dinner, I had the chance to get inside their heads. I wanted a better understanding of where their minds were in those hallway conversations they have with sales reps between surgeries.

Specifically, given the data about how vastly superior my clients' tech was, why would surgeons continue conducting surgery the older, less effective way that put their patients at higher risk with less reliable outcomes?

After going back and forth with one rather exuberant, narcissistic surgeon, I decided to use some of my "voodoo" as one client so eloquently called NeuroSelling to tap into their emotional network. When he finally took a breath, I saw my opening and said, "Look, if I were your patient, I wouldn't let you put me on your operating table. You are choosing to ignore solid medical evidence in favor of older methods proven to be less effective and more harmful to me as a patient. I wouldn't be comfortable having you operate on me or someone I love."

My goal wasn't to make him mad. I purposely played devil's advocate because I knew that if I could spike his stress hormone, cortisol, I'd get an emotional response (instead of a rational, thought-out answer). Finally, one of the surgeons blurted out the heart of the matter: They didn't fully understand the technology, and they weren't comfortable using it.

Put another way: "I pride myself on knowing exactly what I'm doing, having the answers, and being the expert, so I'm going to stick with what I know. This product is new. It's different. I don't want to be the guinea pig in my circle of peers. I don't want to try something new and wind up looking like a fool."

From their point of view, you must appreciate where they're coming from. A stranger is standing in front of them, trying to sell them something, quoting medical studies in journals they may or may not have even heard of. They expect the salesperson to cherry-pick the "facts" that will land them the sale. History and experience have also taught them that for every medical study touting the prowess of their device, there may be a dozen that find contrary results. It's not that these medical professionals don't practice evidence-based medicine. It's that they don't trust the word of yet another medical salesperson trying to sell them something.

Yes, even high-IQ, logical, rational people like doctors and surgeons still function like Grug Crood. **New is bad. New is risky**.

The Universal Fear of Change

The neuropsychological basis for this universal resistance to change is what researchers call "anticipatory anxiety." Studies using fMRI scanning have shown that when people contemplate change, their brain's fear center (the amygdala) activates more strongly when imagining potential negative outcomes than the brain's reward center activates when imagining positive outcomes.[14] In other words, we're wired to give more weight to what might go wrong than what might go right.

This explains why even change that offers clear benefits faces resistance. It isn't irrational; it's the brain's ancient risk-management system at work. For sales professionals, this means that pointing out the positive outcomes of changing to your solution simply isn't enough—you must actively minimize or neutralize the perceived risks of making that change.

That's really the case I'm making in this book. Until your customers feel their current way of doing things (status quo) becomes riskier to their self-preservation than the new alternative you're proposing, they're going to stick with what they know . . . and so would you. There you are, a perfect stranger, telling them it's safe to venture away from what they know. Of course, you're going to tell them that—that's how you make money.

"You don't know me from Adam, but here's something new and different; you should buy it, and by the way, I make my income from convincing you to do that—but trust me, that doesn't affect my recommendation!"

Our survival instincts tell us to stay where we know it's safe until we believe it's safe to do otherwise or, at minimum, less safe than where we are.

Telling them about the features and benefits of your product doesn't persuade strangers because they've been conditioned to *not trust the source*.

No one takes the *National Enquirer* or *Weekly World News* seriously. They're tabloids spouting the latest Elvis sighting or the most recent Bat Boy capture. We know we can't trust the source, regardless of how many "experts" the tabloid article cites. (I'm not saying that surgeons necessarily put medical reps in the same category as supermarket tabloids, but follow my analogy here.)

It doesn't matter what features your product has, how wonderful your service is, how superior your technology is, how many studies you cite, or what benefits you promise—you are a new person, and they believe you're going to look out for #1 (because that's certainly what they're doing) therefore, you likely cannot be trusted. We trust people and sources who've proven to be trustworthy. By default, everyone outside of that circle of trust gets put in the box of "not trusted."

So how do you get into the circle of trust?

THE NEUROSCIENCE OF TRUST

"If people like you, they'll listen to you, but if they trust you, they'll do business with you."

—Zig Ziglar

I ADMIT IT: I'd procrastinated.

When I boarded the plane, I was days behind on some of the administrative tasks that are my least favorite part of being CEO. Thankfully, I'd been upgraded, so I'd have plenty of room to work. I already had my laptop out and started booting up when a mother, father, and son boarded the plane. The mother and son began settling down in the row ahead of me. It looked like the father was going to be sitting beside me.

Now, I'm a sociable guy, but I had a lot of work to do and was really looking forward to knocking it out on this two-hour flight. I stared hard at my screen as I began feverishly typing, hoping I looked too busy to bother. Well, the dreaded fifteen or twenty minutes from the time the boarding door closes until that glorious "ding" that tells us we are above ten thousand feet was the window my new seatmate used to "get to know me." I couldn't even use the "headphones in, even though they aren't connected to anything" trick as I had left them in my bag in the overhead bin.

The moment we began pushing back from the gate, he immediately started chitchatting. I said to myself, *Well, I guess this is how it's going to be.* So, I turned toward him in my seat, gave him a warm smile, and picked up the conversation. I've learned that my real purpose in life is being available for these very conversations. I never know why, but usually, there is a divine reason meant for either me to learn something from someone else or to add value to that person in some meaningful way, so my initial frustration with not getting work done quickly gave way to my past experiences in human connection.

When he asked me what kind of work I did, I started with the same story I tell everyone: "Well, what I do really has to do with why I do it. I'm an old farm boy. I grew up on a hundred-acre farm that my papaw bought…"

After I'd shared my story, I invited him to tell me about himself. He started talking about his life, and we spent the rest of the flight swapping stories and ideas.

When we began our descent into Denver, he asked, "So what does Braintrust actually *do*?"

I said, "We teach people how to communicate with their customers in a more impactful way—from a place of "why" versus "what" based on the science of human behavior."

"Oh?" he responded. "What's that look like?"

"Well, you, for example: I understand where you're coming from. I understand your 'why.' You grew up on a thousand-acre cattle ranch in Colorado, learning life and business lessons from your father and grandfather. They taught you about honesty and integrity, like the time that you watched your grandpa have a chance to keep an extra hundred dollars at the stockyards if he'd have just kept his mouth shut—but he didn't. He spoke up, pointed out the old man's mistake, and paid him the extra hundred dollars. When you started your wealth management firm twelve years ago, you wanted to do business the way you saw your grandpa do business. And that's why it's important to you today that you do the right thing, even when you think nobody's watching."

My airplane buddy sat there, amazed. "Holy cow! How did you get all that from my story?"

I replied, "I told you my 'Why Story.' You instinctively opened up and told me yours."

He said, "You know, I've never told anybody many of the details I shared with you in my story. Not even my wife," he said, gesturing to the seat in front of him.

"Why not?" I asked.

"You know? I don't know, but that's a great point."

On Sunday morning the next weekend, I woke up to see a text from him: "Can't thank you enough for that plane ride. Last night, I shared my Why Story at our company party. Not a dry eye in the house. Changed the whole feel. THANK YOU."

Now, when he asked what I did for a living, why didn't I tell him about Braintrust? Why not give him the standard "Oh, I'm a [insert job title here]"? Why did I answer his question by telling him about my "why"? Why did I *not* tell him how I made my money until the second time he asked?

Because he didn't really care what I did. He was just making polite conversation. Small talk that you expect from your seat buddy.

It wasn't until I'd opened up about myself and shared a deeply personal story—that is to say, something that made me vulnerable—that he felt a real connection to me as a person. Then, without prompting, he shared something deeply personal to him. Something, in fact, that he'd never even shared with his own wife of many years sitting right in front of him!

Once he saw me, not as a stranger, but as a relatable person, he wanted to genuinely know what I did.

This time, instead of asking out of expected politeness, he asked out of genuine interest. When we were landing, he volunteered to exchange phone numbers so we could keep in touch. That's the power of a Why Story.

Even though he was clearly someone we could help, and I was the person with something to potentially sell the minute I realized that, he was the one asking if we could continue the conversation.

The Trust Crisis in Modern Business

This interaction on an airplane wasn't just a pleasant exchange—it was a powerful demonstration of how authentic connection creates trust in a world increasingly defined by its absence. We are currently experiencing what many researchers describe as a "trust recession" in business relationships.

The 2023 Edelman Trust Barometer reports that only 51 percent of respondents trust businesses to do what is right, with that number plummeting to just 43 percent for B2B salespeople specifically.[1] Research from the Edelman Trust Barometer reveals that global organizational trust has eroded significantly over the past two decades, a trend accelerated by economic uncertainty, shifts to remote work, and rising public scrutiny of corporate behavior.[2]

This trust deficit creates real business consequences. Recent Salesforce research found that 88 percent of customers say trust is a critical factor in their buying decisions, and trusted companies earn stronger loyalty, retention, and wallet share compared to competitors.[3] In fact, PwC research shows that even a single negative experience causes 32 percent of customers to stop doing business with a brand they previously loved—demonstrating just how fragile customer trust can be.[4]

The most concerning aspect of this trend is that conventional approaches to addressing it often make the problem worse. McKinsey research found that when sales teams respond to customer uncertainty by simply increasing outreach frequency—without adding meaningful value—they run the risk of creating what I call 'trust fatigue,' which can inadvertently accelerate customer disengagement.[5]

In this environment, understanding the neuroscience of trust—how it's created, maintained, and potentially damaged—becomes an essential business skill, not just an interpersonal nicety.

The Science of Connection

Grug Crood doesn't open up to strangers.

The idea of being vulnerable with someone not "friend" is foreign to him.

And vice versa.

When I displayed vulnerability—when I shared something that was intimately important to me—my newfound airplane friend's caveman brain said, "Oh! A story. Let's listen." His mirror neurons activated. Then, once he realized the depth of the story I was sharing, his caveman brain went, "Oh! Vulnerability. Then this guy isn't a foe—foes don't share personal details like this. Therefore, he must be a friend." His "trust chemical" oxytocin surged.

It's not manipulation. It's learning how to connect and communicate with other human beings in a day and age when our caveman brains are more hindrance than help.

That's why Christina relating her intimately personal experience in the leadership development program created such a profound moment and connection.

The Neurochemistry of Authentic Connection

At its most fundamental level, trust is a neurochemical state. When we experience trust, our brains produce a cascade of neurochemicals that create both emotional and physiological changes. The primary driver of this process is oxytocin, often called the "trust hormone" or "bonding hormone."[6]

Research by Dr. Paul Zak, Director of the Center for Neuroeconomics Studies, has demonstrated that oxytocin is the chemical foundation of trust relationships. When released, this neuropeptide reduces activation in brain regions associated with fear and social anxiety (particularly the amygdala) while enhancing activity in regions associated with empathy and social cognition (including the anterior cingulate cortex).[7]

What makes oxytocin particularly fascinating is how it creates positive feedback loops. In laboratory studies, higher oxytocin levels have been shown to increase a person's willingness to trust others, which in turn stimulates oxytocin release in the recipient of that trust, creating a mutually reinforcing cycle.[8]

This neurochemical exchange explains why authentic connection feels so rewarding. When my airplane seatmate and I shared our personal stories, we were literally changing each other's brain chemistry in real time. His willingness to reciprocate my vulnerability wasn't just politeness—it was a neurobiological response.

The challenge for sales professionals is that oxytocin release requires certain triggers, most notably:

- Perceived vulnerability – When someone reveals something that places them at potential social risk, it signals authenticity
- Empathic resonance – When we feel genuinely understood by another person
- Intentional attention – When someone demonstrates they're fully present and focused on us

These triggers are precisely what typical sales tactics, like rehearsed rapport-building and feature-focused pitches, fail to activate. This failure to engage the oxytocin system explains why conventional approaches often leave customers feeling manipulated rather than connected.

By being vulnerable—and let me emphasize, purposefully vulnerable—you speak directly to the limbic and root brains. Popular researcher and author Brené Brown says it like this in her book *Daring Greatly*:

> *Vulnerability is the birthplace of love, belonging, joy, courage, empathy, and creativity. It is the source of hope, empathy, accountability, and authenticity. If we want greater clarity in our purpose or deeper and more meaningful spiritual lives, vulnerability is the path.*[9]

The most important thing is to speak from the heart. You can't be vulnerable and cagey at the same time. Your listener can tell if you're being insincere or inauthentic. *To create a real connection, you have to be fully authentic and be willing to show it.*

You know why Christina is so passionate about that leadership course. You can see for yourself that I'm carrying on Papaw's legacy of storytelling from my personal "why" story you read earlier.

So, when I say to you, "That's why I do what I do; tell me why you do what you do," I rarely get the other person's resume. They don't say, "My company works with—" They recognize that the mood has changed. I just shared an authentic story with just the right amount of vulnerability about how I grew up and why, thanks to my Papaw, I believe what I believe through what he taught me. That sets the tone for what I'm expecting out of them.

Your Why connects more people to what you do than your What and How. The Why is your passion and reason . . . share it and see how people reciprocate and connect—every time.

In Part II: The Field, we will introduce you to a road map of stories that, if built the right way and delivered in the right order, will drive trust faster and create an urgency for the prospect to change. The first story you will learn to create is your personal connection "why" story. The moment that prospect or customer believes that you're believable, there's a connection, and that connection leads to trust.

The Chemistry of Trust vs. Distrust

While trust creates a positive neurochemical state, distrust triggers a markedly different brain response. Neuroscientist Dr. Antonio Damasio's research using functional MRI has shown that when we experience distrust, the brain's threat detection system activates, releasing stress hormones like cortisol and adrenaline.[10]

This stress response creates several significant effects:

- Neural inhibition – High cortisol levels inhibit activity in the prefrontal cortex, impairing analytical thinking and decision-making

- Perceptual narrowing – The brain's attention narrows to focus on potential threats, reducing receptivity to new information

- Memory encoding changes – Information received during high-stress states is tagged differently in memory, associated with caution and scrutiny

Research published in Harvard Business Review highlights that environments lacking psychological safety—notably marked by distrust—significantly impair cognitive function and creativity, undermining decision quality and problem-solving capacity[11] For salespeople, this means that when customers experience distrust, their capacity to recognize the value of your solution dramatically diminishes.

Most concerning is how quickly distrust can develop. Studies using electroencephalography (EEG) have shown that the brain makes initial trust/distrust judgments within the first hundred milliseconds of an interaction—far faster than conscious thought can occur.[12] These snap judgments create a frame through which all subsequent information is filtered.

One study from Northwestern University's Kellogg School of Management found that overcoming initial distrust requires approximately nine to twelve positive interactions to counteract a single negative first impression.[13] This asymmetry explains why so many sales relationships struggle to recover from poor starts.

The key takeaway: Trust isn't just a psychological state—it's a neurobiological condition that fundamentally changes how information is processed, decisions are made, and memories are formed.

You may believe that you couldn't possibly create a Personal Why Story. We've heard thousands of people say those very words over the past decade. But everybody has a story. And I mean everybody. Everybody comes from somewhere. Everybody has a reason for why they do what they do. We just have to help you find yours and then teach you how to communicate it the right way at the right time, just like Matt Rogers . . .

The Power of the Personal Narrative

Early on the first day of a two-day NeuroSelling workshop, I could tell Matt was the ringleader of the salespeople in the group. He was built like a defensive lineman, and, as I learned, he had, in fact, been a football player in college. He also had a gregarious personality and natural charisma about him (which helped when he'd been on *American Idol*—the guy was just all-around impressive).

But in the two-day workshop, I could see he was struggling with the part of the training where we help people write their "Why Story." What's a Why Story, you ask? Well, think back to Chapter One where I told you my Papaw story, then ended with why teaching neuroscience and narrative-based selling is personally close to my heart. I essentially explained what I do by embedding it inside a memorable story focused on why I do it.

Of course, it's not easy. But when you get it right, wow, is it effective!

Seeing that Matt was having some challenge with the exercise, during a break, I pulled him out into the hall and said, "What's up, big guy? What's got you stumped?"

He said, "I don't know what to write about. I just don't have a motivating story, and I just don't see why any customer would want to hear about it even if I did."

I said, "Well, tell me a little about how you grew up."

"I didn't have a great childhood. My dad wasn't really around, so I was basically raised by a single mom."

"Okay," I prodded, "tell me about her."

"My mom? Gosh, she's amazing. I don't know how she did it. She worked two jobs to keep me and my brother in clothes. And our clothes didn't come cheap because we were both big guys."

He went on for a few minutes, then I asked, "So what do you feel like your mom taught you? Any examples you can remember?"

"Yeah, she taught me a lot about sacrifice, serving others and thinking about others more than yourself. Even though we ourselves were poor, every Saturday morning, we had to go down to the homeless shelter and serve food to others my mom considered far worse off than us. That was her thing. I'll never forget one time—I was probably thirteen years old—and I wore a size fourteen shoe already. I had been saving up my money from doing chores and odd jobs so that I could finally have a pair of the new Nikes. My mom had been saving up as well, as she wanted to match what I had saved to help me out. I had my eyes on these shoes for months and months. After a lot of hard work and saving, I finally got that pair of Nikes I wanted. I was so happy. I'd never had anything nice like that before. We didn't have money, so this was a big deal.

"Well, that Saturday, we went to serve at the homeless shelter. I remember it was a really chilly morning. On the way in, we walked past a homeless guy sitting on the side of the street. My momma stopped to talk to him like she did with everybody and invited him inside to get a hot meal. Then she looked down at his shoes. They were too small, so he'd cut the toes out so he could wear them. His toes were just sticking out and I remember thinking how cold that must have felt."

Matt stopped here, choking back tears as he recounted the rest of his story.

"She told me, 'Matt, take off your shoes and give them to this man.' I was bigger than my momma, but she was one of those women you just don't argue with. So I gave him my shoes, and then she made me put on his. Otherwise, I'd have gone barefoot.

"I was devastated. I'm watching this guy walk around the shelter and get his food tray, all while wearing the Nikes I've been saving up to buy for nearly a year. When we got ready to leave, I thought she was going to make him give them back, but she didn't. We just got in the car and drove home, me with my big, size fourteen feet sticking out of these ratty, stinking shoes, with my toes now freezing.

"And then she said this to me: 'Matt, I don't want you to ever forget this: You will never know what someone's life is like until you walk in their shoes.' And as sad and disappointed as I was in that moment, you can bet I've never forgotten that." (She later surprised him with a replacement pair, but not before that lesson was fully embedded deep within Matt's memory bank!)

By the end of his story, he was choked up, I was choked up, and other people were passing us in the hall wondering what was going on, but I said, "Matt, why would you say you don't have a Why Story? Get your butt in that room and start writing!"

He went on to craft a personal Why Story so engaging the entire workshop gave him a standing ovation after he told it at the end of the class. And here's the thing, he volunteered to do so. As I often quote from Mark Twain, "The two most important days in your life are the day you were born and the day you discover why." Matt certainly did just that.

Six months later, he called me from his cell phone.

"Jeff, I've got some good news. At the workshop, I was number 183 out of 300 salespeople in our company."

Not sure where this was going, I said, "Okay?"

"Well, you know the story I didn't have, that I really did have?"

Of course, I remembered. Who could forget a story like that?!

"Well, next week, I'm getting recognized onstage as the top salesperson in my division and one of the top in the entire company! And it's all because of my Why Story."

I said, "Alright! Matt, that's awesome!"

Matt said, "When I go on a sales call, I barely make it past my story when they say, 'I don't even care what you're selling, I'm buying!' It's been unbelievable."

The Neural Mechanisms of Narrative Persuasion

Matt's transformation from the bottom half of his company to top performer wasn't accidental or mysterious—it's a predictable outcome of the neural processes triggered by personal narrative. Narrative persuasion operates through several key mechanisms that conventional sales approaches can't access.

Neural Coupling and Synchrony

Research by Green and Brock demonstrated that when individuals become emotionally "transported" into a narrative, they experience significantly greater persuasion compared to when the same information is presented through non-story formats.[14] Using functional MRI, Hasson's team observed that when someone tells a compelling personal story, the same brain regions activate in both the storyteller and the listener, creating a remarkable form of brain-to-brain synchrony.

This shared neural activation doesn't occur during typical business presentations. It specifically emerges during narrative communication that includes emotional components and sensory details. When Matt shared his story about giving away his Nike shoes, his customers' brains were literally mirroring his own emotional experience—creating a level of empathy that data and features cannot produce.

Narrative Transportation

A second mechanism at work is what communication researchers call "narrative transportation"—the experience of being so absorbed in a story that you temporarily leave your current reality.[15] This state, characterized by reduced critical thinking and increased emotional engagement, creates an ideal condition for persuasion.

When people enter a transported state, several important changes occur:

1. Reduced counterarguing – Critical analysis of the message decreases
2. Increased emotional response – Empathic connection with the narrator intensifies
3. Identity simulation – The brain processes the story as if the listener were experiencing it directly

Research from The Ohio State University found that narrative transportation increases persuasive impact compared to non-narrative messages.[16] This explains why Matt's customers were saying, "I don't even care what you're selling, I'm buying!" after hearing his story—they were making decisions from a transported state where emotional resonance trumped analytical evaluation.

Take a wild guess at what Matt sold. What came to mind? Life-saving pharmaceuticals? Life-changing coaching? Paradigm-shifting consulting? World-changing technology? Would you have guessed outsourced payroll and HR?

What is it about Matt's narrative that took him from the bottom half of his sales organization to the top? He starts each sales conversation not with facts and figures aimed at the logical neocortex but by first connecting on a subconscious level. He establishes personal trust.

All this goes back to how our brain processes information and makes decisions. We make decisions that make us feel safe. We feel safe with information we trust. We trust information that comes from trustworthy people. We trust people we connect with.

The Vulnerability Paradox in Professional Settings

One of the most counterintuitive aspects of trust-building is what researchers call the "vulnerability paradox"—the phenomenon wherein appropriate vulnerability in professional settings increases rather than decreases perceived competence and authority.[17]

This runs contrary to traditional business wisdom, which often counsels professionals to project unwavering confidence and certainty. However, research from Harvard Business School has found that leaders who demonstrate strategic vulnerability are rated as 27 percent more effective and 41 percent more likely to inspire discretionary effort from their teams than those who maintain a façade of perfect competence.[18]

The neurobiological basis for this paradox involves what social neuroscientists call "costly signaling." By showing vulnerability—which would be disadvantageous if you had harmful intentions—you trigger trust circuits in the observer's brain that recognize this as evidence of benign intent.[19]

However, this paradox operates within tight boundaries. Research by social psychologist Brené Brown has identified that vulnerability becomes counterproductive when it:

- Is inappropriate for the relationship stage (too much too soon) Demonstrates lack of boundaries (oversharing)
- Appears performative rather than authentic
- Shows no reciprocity (only one party is vulnerable)[20]

When appropriately calibrated, vulnerability in professional settings provides what anthropologists call "trust verification"—evidence that you are who you claim to be.[21] For sales professionals, this creates a powerful differentiator in a world where customers are constantly filtering for authenticity.

Ditch Rapport. Build Trust.

When I say personal connection, most sales professionals hear "rapport building." **Repeat after me**: *Rapport building does not mean you've established personal trust.*

In fact, you've probably never been taught how to build a personal connection in a business environment. Think about it: How much time in your career have you spent in classes, trainings, workshops, reading, and learning how to build genuine personal connections in a business environment? Most of the sales techniques you've probably heard are around "rapport building," a quick-and-easy, superficial, artificial, transactional, self-serving gimmick to trick people into trusting you. (Even if your intentions were pure, your motive is still self-serving.)

The Failure of Conventional Rapport Building

Traditional rapport-building techniques fail to create authentic connection for neurological reasons that are now well-documented. Research from the University of Southern California's Brain and Creativity Institute has shown that the brain has specialized circuits for detecting social authenticity—what researchers call the "authenticity detection network."[22]

This network, which includes regions like the anterior insula and medial prefrontal cortex, activates when we encounter behavior that appears calculated rather than genuine. When these regions fire, they trigger what neuroscientists call a "social warning signal" that promotes skepticism and protective behaviors.[23]

This is precisely what happens during conventional rapport-building attempts. When a salesperson uses techniques like:

- Finding commonalities based on office decor ("I see you like golf too!")
- Mirroring body language in an obvious way
- Using forced familiarity ("How's the family?")
- Deploying obviously rehearsed small talk

The customer's authenticity detection network activates, creating the opposite of the intended effect—increased vigilance rather than increased trust.

The one I tend to see over and over again looks a little something like this: You walk into a sales meeting, look around, and see the customer has a picture of themselves on a sailboat.

"Why, Mr. Smith, I see you have a picture of a sailboat on your wall. You like sailing? What a coincidence—I drink water! Looks like we already have something in common!"

That sounds amateurish when you read it in black and white, but I cannot count the number of professional salespeople selling multimillion-dollar products who do this type of rapport-building attempt, maybe not word for word, but believe it or not, this is likely how it comes across to your customer. Maybe it's a picture of their spouse, children, pet, or hobby. Regardless, the salesperson tries to find some way to identify with the customer: "Oh, you have kids? Me, too! Man, aren't they great? Okay, let's talk about your software needs!" Does it feel like that approach creates a personal connection? Or does it sound like an impersonal attempt to get their business? The hard thing for us to accept is that the customer knows exactly what you are doing. They've seen that movie a thousand times before with a thousand other salespeople. Here's the thing: no matter how genuine you come across using this type of technique, you can only create, at best, a superficial level of trust. Not the type of trust that drops the shields.

As soon as their caveman Grug Crood brain sees you as a transactional "Can I get you in a new car today?" type, it thinks, *Okay, here's another salesperson trying to get our money. Enemy alert—shields up!*" Their cortisol spikes. Instantly, their brain goes into defense mode and begins filtering everything through the lens of risk and skepticism.

You know what doesn't spike?

Oxytocin.

The Trust Chemical

Oxytocin is a neurochemical/neuropeptide that our brain produces that was originally discovered by studying its effects after a mother gives birth, binding the mother and child together[24] as well as fathers.[25] Researchers have nicknamed it "the bonding hormone" or "the love hormone," but we have determined a better name is actually "the trust chemical."

The oxytocin in our brain rises when we interact with our "in-group": individuals we've identified as friend, not foe. *Oxytocin solidifies our connection with others and is why we feel protective of them.* We would do things for our in-group that we would never consider doing for strangers.

We even experience a rise in oxytocin when we interact with our pets.[26]

Oxytocin also reduces our feelings of fear and anxiety,[27] actually counterbalancing the effects of adrenaline and cortisol triggered by our brain's fight-or-flight instincts. We've discussed how our customer's defenses are already triggered by a sales meeting. Encouraging oxytocin can help moderate those instincts.[28] Perhaps even more importantly, **higher levels of oxytocin mean higher levels of perceived trustworthiness.**[29] At high enough levels, research suggests that it can overcome breaches of trust. That is, even if you do something to me, if I have enough oxytocin, I'll still trust you despite your actions.[30]

The Neuroscience of Trust Building

Recent advances in neuroscience have revealed precisely how oxytocin functions in trust building. Dr. Paul Zak's research team at Claremont Graduate University has identified specific behaviors that reliably trigger oxytocin release and build trust:[31]

- Recognition – Acknowledging someone's unique qualities and contributions
- Intentional attention – Giving someone your complete, undivided focus

- Shared vulnerability – Revealing appropriate personal information
- Demonstrated empathy – Showing genuine understanding of someone's situation
- Natural synchrony – Matching communication style and energy authentically

These trust-building behaviors operate on a neurological level, not just a psychological one. When deployed effectively, they trigger oxytocin release in both parties, creating what Zak calls a "positive feedback loop of trust" that can rapidly transform the relationship dynamic.[32]

This explains why Matt's Why Story transformed his sales results—he was literally changing his customers' brain chemistry. Simply put, when two human beings interact, and there is the perception of care, connection, empathy and trust, oxytocin levels are high in both participants.

Harvard Business School professor Amy Cuddy has been studying first impressions for more than fifteen years and has discovered patterns in these interactions.

In her book *Presence*, Cuddy says that people quickly answer two questions when they first meet you:

- Can I trust this person?
- Can I respect this person?[33]

Interestingly, Cuddy says that most people, especially in a professional context, mistakenly believe that competence (credibility) is the more important factor. After all, they want to prove that they are smart and talented enough to handle your business, right?

But, in fact, what she found was that personal trustworthiness is the most important factor in how people evaluate you and when they feel that way about you, oxytocin goes up in both parties!

Cortisol, aka "the stress hormone," is the anti-oxytocin. When my caveman brain is in "friend-or-foe, fight-or-flight?!" mode, it overrides whatever oxytocin I had as my stress is triggered. And until my brain moves you from the category of foe to friend, my cortisol will remain high.

So how can you get my oxytocin flowin'?

With what I call the elements within the "Periodic Table of Trust."

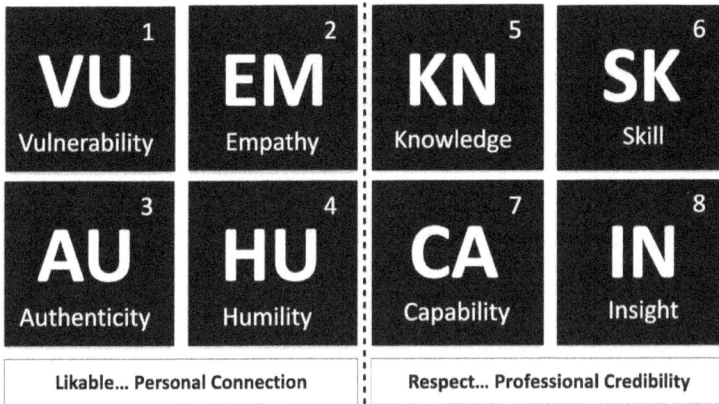

VU ¹ Vulnerability	**EM** ² Empathy	**KN** ⁵ Knowledge	**SK** ⁶ Skill
AU ³ Authenticity	**HU** ⁴ Humility	**CA** ⁷ Capability	**IN** ⁸ Insight
Likable... Personal Connection		**Respect... Professional Credibility**	

Personal Trust Elements

Vulnerability (VU) – Sharing appropriate personal information that creates connection Authenticity (AU) – Behaving in alignment with your stated values and beliefs Humility (HU) – Acknowledging limitations and being open to others' expertise Empathy (EM) – Demonstrating genuine understanding of the other person's perspective.

Professional Trust Elements

Professional trust is built on a foundation of Knowledge (KN), which reflects a deep understanding of subject matter and specialized expertise relevant to the customer's needs. Skill (SK) is the consistent application of learned abilities to meet expectations, deliver on promises, and execute with precision. Capabilities (CA) represent the ability to solve complex problems, produce measurable results, and leverage valuable resources or networks that enhance outcomes. Insights (IN) provide strategic perspective, foresight, and contextual awareness that help customers make more informed decisions and move confidently toward their goals. When demonstrated consistently, these four elements form the backbone of lasting professional trust.

Importantly, the sequence matters. Personal trust elements must precede professional trust elements in the trust-building process. When this sequence is reversed, professional credentials actually increase scrutiny rather than decrease it.

The Sales Trust Matrix: Moving from Acquaintance to Advocate

The culmination of the research on trust building is what we call the "Trust Matrix," a framework for understanding how relationships evolve from initial contact to deep partnership. This matrix maps the journey customers take as they move through increasing levels of trust, and how salespeople can facilitate this progression.

The Four Levels of the Trust Matrix

Level 1: Acquaintance Trust

Every relationship begins with a first impression. At this level, trust is tentative and primarily driven by emotional heuristics—quick judgments based on tone, appearance, perceived intent, and nonverbal cues. This aligns with research by Dr. Alexander Todorov at Princeton, who found that people form judgments of trustworthiness within milliseconds based on facial expression and body language alone.[34]

At this stage, the customer's brain is operating largely out of the amygdala and insula, scanning for social threats and signals of safety. If your approach feels safe, warm, and genuine, the customer may allow the interaction to continue—but skepticism remains high.

The goal at this level isn't to sell. It's to establish a safe emotional environment where the customer's brain lowers its guard enough to even consider hearing what you have to say.

Salespeople who skip or rush through this level risk activating social threat networks that shut down trust before it can even begin.

Level 2: Knowledge Trust

As customers receive valuable information and insights, they begin to build what's known as competence-based trust. This kind of trust is rooted in the belief that the other person knows what they're talking about and can be relied upon for sound advice. At this level, people tend to think more critically and analytically while still keeping their emotional guard up. Customers may start to share some details about their situation, but they're still careful about revealing too much.

Level 3: Relationship Trust

When consistent, authentic interaction combines with demonstrated value, customers transition to what behavioral economists call "integrity-based trust." Research from Dr. Diana Tamir at Princeton University shows that this level of trust activates the brain's reward circuitry, creating positive associations with the relationship itself.[35] At this stage, customers become more forthcoming with challenges and strategic objectives.

Level 4: Advocacy Trust

The highest level occurs when customers not only trust you with their business but actively promote you to others. Neuroimaging studies show that this level of trust actually changes how the receiver's brain encodes information about you, moving your identity representation into brain regions associated with their in-group (medial prefrontal cortex) rather than those for transactional relationships (temporoparietal junction). This neurological shift explains why true advocates defend your relationship even when problems occur.

Moving customers through this matrix isn't just good relationship building—it's good neuroscience. Each level activates different neural networks, gradually shifting how the customer's brain processes information about you and your solutions.

Trust in Action: The 3-Step Trust Process

How do you put this neurological understanding of trust into practical action? Research has identified a three-step process that reliably builds trust through the proper sequence of neurochemical triggers:

Step 1: Connection before Content

Begin by activating oxytocin through authentic personal connection before attempting to share professional content. Research shows that information received after oxytocin activation is processed differently in the brain—with greater emotional salience and reduced skepticism.

Step 2: Why before What

Share your personal motivation and purpose before discussing your products or services. Neuroimaging research has found that "why" information activates the limbic system (emotional brain), while "what" information primarily engages the neocortex (analytical brain). When these are sequenced properly, the emotional engagement creates a frame for the analytical content.

Step 3: Vulnerability before Credibility

It may seem counterintuitive, but leading with appropriate vulnerability can actually strengthen your professional credibility. When you share something genuine or personal first, people tend to be more open and receptive to what comes next. In fact, your credentials and expertise often carry more weight when they follow a moment of vulnerability because trust has already begun to form.

Think back to my airplane conversation. I shared vulnerability first (my personal story), which created connection before I shared content (what Braintrust does). I began with why (my Papaw's influence) before what (our work with communication). And I demonstrated appropriate vulnerability before establishing credibility. The result was a genuine connection that transcended typical business interactions.

The Trust Revolution: Where We Go From Here

We are witnessing what social scientists are calling a "trust revolution" in business relationships. As traditional advantages like information asymmetry and product differentiation continue to erode, trust is emerging as the primary competitive differentiator in sales relationships.

As business decision-makers continue to site trust as the key attribute in vendor selection, neuroscience research continues to deepen our understanding of how trust operates at the brain level, providing increasingly sophisticated frameworks for building authentic connection.

The future of sales belongs to those who understand not just the psychology of trust but its neurobiology—the complex interplay of chemicals, neural networks, and evolutionary wiring that determines who we connect with and why.

In the chapters that follow, we'll translate this scientific understanding into practical frameworks, specific language patterns, and step-by-step processes for building trust-based customer conversations. You'll learn how to craft your own Why Story, develop the additional NeuroSelling narratives, and remove the neurological barriers to change.

But first, let's summarize the key principles from this chapter:

- Trust is a neurochemical state primarily driven by oxytocin. Authentic connection through appropriate vulnerability is the most reliable trigger for oxytocin release

- The brain processes information differently in high-trust vs. low-trust states

- Trust building follows a specific sequence: connection before content, why before what, vulnerability before credibility

- Digital environments create unique challenges for trust building that require specific adaptations

As we move from understanding the science to implementing the strategy, remember Matt Rogers and his remarkable transformation. He didn't change what he was selling. He didn't acquire new technical skills. He simply learned to communicate in a way that aligned with how the human brain naturally builds trust.

That same opportunity is available to you, and in the coming chapters, we'll show you exactly how to seize it.

THE PHYSIOLOGY OF CHANGE RESISTANCE

"Resistance to change is proportional to how much the future might be altered by any given act."
—Stephen King

I'D LIKE YOU to pause for a moment and think about the life you've created. From the job you have to the home you live in to the relationships you've developed—from your spouse or significant other to your closest friends. Now, just for illustration purposes, I'd like for you to imagine that in the span of just a few months, your house catches fire, your spouse asks for a divorce, and your boss tells you that your services are no longer needed. This scenario, as unimaginable as it may seem, was the unfortunate case for James. He certainly wasn't looking for this to happen, didn't *choose* for any of it to happen, yet like it or not, his life had dramatically and suddenly changed.

In our live workshops, I frequently ask the group, "By a show of hands, how many of you like change?" Inevitably, about half the room will raise their hands.

Then I rephrase the question: "How many of you like change that was chosen for you or forced upon you that you didn't initiate?" You start to see the hands slowly go down in the room. Nobody likes feeling forced to change.

What is it about change? Why are human beings resistant to it? Why do we especially resist change when it's not our idea, even when it's a rationally better option?

Now, in James's case, those clearly weren't better options, but he certainly didn't choose for them to happen. As you might imagine, James feels that change is bad. In fact, most, if not all, humans feel the same way.

The Neurobiology of Change Resistance

At its core, resistance to change is hardwired into our neurological systems. Research by neuroscientists David Rock and Jeffrey Schwartz has demonstrated that the brain expends significant energy, maintaining a state called "homeostasis"—the tendency to seek and maintain a state of equilibrium.[1] When faced with change, the brain's threat response activates, primarily in the amygdala, triggering what is commonly known as the "fight, flight, or freeze" response.

This neurobiological reaction isn't just psychological; it manifests in measurable physiological changes:

- Increased heart rate and blood pressure
- Elevated cortisol (stress hormone) levels
- Reduced blood flow to the prefrontal cortex (responsible for rational thinking)
- Heightened sensory alertness
- Muscle tension and preparation for action

What's fascinating is that these physiological responses occur whether the change is objectively beneficial or harmful. The brain's primary concern is predictability, not optimization. Behavioral economists Kahneman and Tversky's research on status quo bias confirms this neurological preference for what's familiar, regardless of the potential benefits from change.[2]

For instance, I once worked with a client several years ago who decided that, instead of having laptops for their sales organization, they were going to roll out tablets for everyone—all two thousand salespeople. The strategy behind the change was sound: better flexibility, more versatility, and the sales team would look more "tech-savvy" to their customers.

Put yourself in the salesperson's shoes. Your status quo is your laptop. You understand it. You know how it works. You've been using it for years. It's been your number-one weapon as you leave the cave each day to kill something and drag it home.

Suddenly, your boss's boss's boss says, "We're taking away your laptop. You have to use an iPad now. It'll be great. Good luck."

If you were an early tech adopter and liked the latest gadget, you would likely already have a personal iPad, so you'd be excited about this change: It's change you welcome or even want because it aligns with how you see or feel about yourself! By definition, though, most people are not early adopters. Most people like to go with something that's proven, especially when it's how they make their living. The vast majority of those two thousand people were unhappy, frustrated, or even angry.

There is a very big difference between the change that we choose versus change that we feel like someone else forces us to do.

The Status Quo Premium

Recent research in the field of behavioral economics has quantified what psychologists have long observed—people place a premium on maintaining their current state. A landmark study by William Samuelson and Richard Zeckhauser found that individuals required potential gains of two to two and a half times the value of their current position to willingly undertake significant change.[3] In other words, a 100 percent improvement isn't enough; most people demand a 200–250 percent improvement before they're willing to leave their comfort zone.

This "status quo premium" explains why even objectively superior products and services often struggle to gain market share from established competitors. It's not just about the relative value; it's about overcoming the inherent resistance to leaving what's familiar.

If you come to me with something that challenges my current status quo, it feels like you're challenging me personally.

You're confronting my beliefs about myself, my work, and my previous decision-making.

You're telling me, "Jeff, what you've been doing all this time is wrong. My way is better. You need to change."

In fact, that was almost literally my message to those surgeons at the conference I previously mentioned. Of course, it triggered them; that was my goal. They had to explain to me, in a moment of emotion, the reason behind their change resistance.

When you challenge someone's status quo either intentionally or unintentionally, the other person's inner Grug Crood speaks up: "Change is bad. Anything new is bad. New ideas are very bad." Their cortisol spikes, and their stress mechanisms start firing because their brain doesn't know if it's safe.

The Neuroscience of Change Resistance

To understand why change is so difficult, we need to understand how the brain processes it. Neuroimaging studies have revealed that contemplating significant change activates the same neural networks as physical pain.[4] This is not metaphorical—the same brain regions that process physical pain (the anterior insula and anterior cingulate cortex) become active when we face unwanted change.

This neural response explains the familiar sensation of discomfort we feel when confronted with altering our routine or adopting new methods. It's not simply psychological resistance; it's a genuine neurological reaction that shares characteristics with our response to physical threat.

The threat of change triggers our self-preservation orientation.

When you walk into a sales conversation, you're asking your potential customer to leave their comfort zone. You're telling them they need to change. Who likes to be told they're wrong? Who likes to be told they need to do something better?

No one. Ever.

Change is not usually fun. All of us already feel like we're overwhelmed as it is, just keeping up with everything we're already doing. Then someone like you comes in and tells me I need to learn a new feature, try out a new service, or take time out to experience a new widget? You're not *just* trying to get me to leave my comfort zone. You're not *just* trying to overcome my self-preservation orientation. You're pouring gasoline on all the fires in my life. I don't want gasoline—I'm trying to put the fires out!

When you approach your customers, it doesn't matter how much better things could be for them if they'd only buy what you're selling. As the great Dr. Henry Cloud once said, *"We change our behavior when the pain of staying the same becomes greater than the pain of change."*

That's such an important concept that it deserves its own block quote reworded to fit today's sales professional:

People will only change when they perceive the risk of maintaining the status quo exceeds the risk of changing.

This principle has been empirically validated through numerous psychological studies. Work by psychologist Kurt Lewin on his "Force Field Analysis" model demonstrated that successful change only occurs when the forces driving change outweigh the forces resisting it.[5] The key insight for sales professionals is that reducing perceived barriers to change is often more effective than increasing the perceived benefits of change.

When speaking with your customers—and especially when speaking to them for the first time—it's incumbent on you as a professional communicator to understand a few specifics around **barriers to change**. We will focus on what I believe to be the top six barriers. These are driven largely by the limbic brain (the feeling brain) and the root brain (the instinctive brain).

You need to be able to articulate a message that puts your customer's mind at ease so they're willing to go along with you on this change journey.

As an aside, when it comes to implementing the concepts throughout this book, you will also potentially be challenged with doing something different with the way you communicate, so naturally, you will likely experience these barriers as well.

Barrier 1: The Anxiety Avalanche Effect

In Gary Keller's book *The ONE Thing*, the founder of the global realty company Keller Williams talked about "the clench":

> *You might say that I started to clench my way to success. I really did. I thought that this might be the way you went through life—with your jaw clenched, your fist clenched, your stomach clenched, and your butt clenched.*[6]

Fortunately, he discovered a better way forward than simply living in a constant state of elevated cortisol, always in fight-or-flight mode. But it's a great visual to keep in mind when you are going into a customer conversation . . . for both of you.

Recent neuroimaging research has provided fascinating insights into what happens during this "anxiety avalanche." When we experience acute stress, the amygdala—our brain's threat detection center—becomes hyperactive while simultaneously reducing activity in the prefrontal cortex, the region responsible for rational thinking, planning, and impulse control.[7]

This physiological response creates what neuroscientists call "cognitive tunneling"—a narrowing of mental focus that prevents us from seeing the bigger picture or evaluating information objectively. In this state, we become hyper-focused on potential threats and significantly less receptive to new ideas or perspectives.

Even if you're an experienced sales veteran pulling down seven figures, you can't fool yourself: Your brain knows that you're walking into an important, high-stress situation. Your body will be under stress. You may have far less stress than when you first began your career, but there's no way for you to completely switch it off. Each of these conversations could be the difference between hitting quota or not. Your self-preservation orientation automatically triggers—the clench.

You have to realize that the same "clench" reaction is true for your customer.

You may have more riding on the outcome of the conversation than they do, but regardless of the context, I guarantee they have heightened anxiety too. I might be used to a parade of salespeople coming through my office or operating room, but just the fact that you're not in my in-group and the knowledge that you're job is to get me to do something differently than I'm currently doing immediately triggers my self-preservation orientation. My adrenaline kicks in, my anxiety rises, and cortisol is spiked, no matter how confident I am or may appear to be. My limbic and root brains know exactly what's going on. You're an unknown; you want me to leave my safety box and charge off into the wild unknown. Nope. Not happening.

How do you overcome this barrier?

It's not by appealing to my logical mind. Mr. Spock might have been the Vulcan aboard Enterprise, but the irrational, fly-by-the-seat-of-his-pants, shoot-from-the-hip human being James T. Kirk was the captain. Presenting things like testimonials, social proof, lab results, clinical trials, and data-driven schematics isn't going to persuade Captain Kirk to lower his shields. Not until he trusts the source of that information. Mr. Spock is the captain's sounding board, but Jim's in charge of the ship.

To counteract the cortisol, you have to get the oxytocin flowin'. That means a narrative embedded with the elements of personal trust to get their mirror neurons firing. I'll teach you a specific way to build and deliver that narrative in Part II: The Field.

Barrier 2: The Desolation of Isolation Effect

Think back to those two thousand reps suddenly forced to switch from laptops to iPads. They got frustrated. Even angry. There was a mass outcry and pushback. Some business managers often chalk these types of reactions up to "people just hate change" and direct their employees to suck it up in so many words. Fortunately, my clients weren't cut from that cloth; they listened in earnest to their sales reps. And you know what they discovered was the real problem?

The vast majority of those reps felt isolated.

With all two thousand salespeople being geographically dispersed, they didn't have an opportunity for watercooler talk. They didn't know that many of the other 1,999 reps felt the same way they did. They intuitively assumed everybody else was successfully making the transition while they alone struggled with making the switch. Each one felt like they alone had been left out in the cold to fend for themselves.

The psychology behind this isolation effect is deeply rooted in our evolutionary history. Humans developed as tribal beings, and social connection has always been essential for survival. Neurobiological research has shown that social isolation activates many of the same neural pathways as physical pain.[8] In fact, some studies suggest that social isolation can be more damaging to health than smoking fifteen cigarettes a day.[9]

When you walk into my office asking me to change, you're asking me to ride even farther into the Wild West. To use another cowboy metaphor, you want to cut me off from the herd. That's how lions and cheetahs catch their prey: by isolating the weak from the pack and killing them individually. I feel safer if I remain with my herd.

(Sure, some customers like being on the bleeding edge of technology and methods, but they're not going to make a major business decision just because they want to be one of the cool kids. Even with them, you need to remember that you're up against their potential feelings of isolation.)

In today's more complex B2B buying environments, there is usually more than one person involved in the buying process. You might think that this would spread the impact of a bad decision across more decision-makers, which, in turn, would decrease the feeling of isolation, but it tends to have an almost multiplication effect. Each person in the process feels the pressure of being the lone person to recommend the change, and the politics within organizations can cause prospects to fear being the only person on their team to advocate for the "risk of change."

Scientists have documented a phenomenon called "pluralistic ignorance," where individuals privately reject a norm but publicly uphold it because they incorrectly believe that most others accept it.[10] This creates a situation where everyone is thinking the same thing ("This change seems risky"), but nobody speaks up because each person believes they're alone in their concern.

How do you work with this barrier to change?

Simple: Make sure I don't feel isolated. Create a personal connection with me so that I feel like I have a partner willing to go with me, not someone who's just there to make the sale and hand me off to technical support. Show me that I won't be the Lone Ranger:

Who else has made the switch? What's been their experience? Do you have any social proof to show me that others have walked this path before? Have you been able to connect the various decision-makers and influencers in my organization in a way that creates a more visible and "team" approach to our buying process? Finally, have you been able to show me how not changing actually puts me and the goals and objectives of my role at risk?

If you switch my perspective and mindset, that resetting can help remove my fear of isolation by helping my subconscious focus on our next change barrier.

Barrier 3: The Fear of Loss Effect

The psychology behind our fear of loss fascinates me. We've lived in a world of scarcity for so many generations of cavemen that it's etched into our collective human subconscious. Starving to death was more than a real fear: It was a reality. Finding a good source of food, water, or shelter was rare. When our ancestors did, they guarded it with their life because it often was a matter of life or death.

Even if that's not the reality for you or me anymore, our brains haven't caught up. We're hardwired to hold on to what we have and literally fear its loss, whatever "it" is. We've immortalized this in the age-old proverb, "One in the hand is worth two in the bush."

When you present me with the option to do something different, you're saying, "Jeff, I know you have done what works for you—even if it really doesn't—but here, try this. You'll love it. Trust me."

You're asking me to give up what I know in exchange for some unknown. I may not like where I am, but I'm afraid of losing what I already have or the predictable results I'm already getting. You're asking me to gamble my job or reputation on whatever you're selling.

The psychological principle at work here is what Kahneman and Tversky identified as "loss aversion" in their Nobel Prize-winning Prospect Theory research. Their studies consistently found that people feel the pain of loss approximately two to two and a half times more intensely than they feel the pleasure of an equivalent gain.[11] This means that a potential 50 percent improvement in outcomes is generally not compelling enough to overcome the fear of a 25 percent loss.

In business contexts, this manifests as what economists call the "endowment effect"—the tendency to overvalue what we already possess simply because we own it.[12] Experiments have shown that people typically demand much more to give up an object they own than they would be willing to pay to acquire the same object.

Since my brain prioritizes not losing what I already have, my natural inclination is to view change as bad or risky, predicting what I might lose; therefore, I prefer the status quo.

Use that as a frame when you engage in customer conversations.

But I'm not a fear monger, and you shouldn't be, either. I've found a way to use this change barrier in a way that feels authentic, not sleazy.

I reframe the customer's problem by introducing a new point of reference: **What do they stand to lose if they *don't* act . . . *don't* change?**

In fact, you spike your customer's cortisol with this angle, not enough to kill their oxytocin, but enough to trigger a slight surge of adrenaline, helping them focus on their real problem (the one you're helping them with).

When their focus goes from the risk of trying something new to the certainty of doing nothing, this barrier to change flips into a motivator: They want to run away from loss as fast as possible!

Barrier 4: The Options Overload Effect

My wife used to own and run a creative meeting venue in Cincinnati called Brainstorm. (Yes, I had influence on the name, and yes, I may be a little too obsessed with the brain.) Nonetheless, she had two locations. One downtown and one in the suburbs. One typical morning, while I was traveling for a keynote talk out of town, she called me in a bit of a panic. She had events going on in both locations on the same day. Both venue hosts called her at 6:00 a.m. and had forgotten their keys and couldn't get in. The caterer who was supposed to deliver breakfast to the downtown location showed up at the other location, causing breakfast chaos to ensue.

On her way to the venue to let her employee in, my wife received a call from a neighbor downtown that a crane putting in new windows on the floor above hers accidentally knocked in the back wall of her venue with the crane. Then, just for fun, the fire marshal called her just a few minutes later to inform her of multiple changes she needed to make to the venue to avoid being out of compliance. This all transpired on the same morning, while I was out of town and she was trying to get our, at the time, seven-year-old daughter, Priya, ready for school.

Have you ever had a day like this or a season like this? Maybe even a year like this where it seems like "When it rains, it pours."

Well, like a good husband who believes "problem-solvers rule the world," when my wife called to tell me about this almost unbelievable cascade of events, naturally, I went into problem-solver, superhero mode. I began to list off in rapid-fire fashion a solution to handle each situation. I could hear my wife's voice start to break on the other end. She had hit the breaking point. As humans, we can only take on so much before we start feeling overwhelmed. **We can only take so much change at one time and so many choices or options for solutions to that change before our brains just shut down.**

Your customers feel the same way.

This principle is well-documented in psychology as "choice overload" or "decision fatigue." Pioneering research by psychologist Barry Schwartz demonstrated that while some choice is good, too many options can paralyze decision-making and reduce satisfaction with our eventual selection.[13] In his landmark book The Paradox of Choice, Schwartz describes how an abundance of options results in:

1. Decision paralysis

2. Lower satisfaction with chosen options

3. Escalation of expectations

4. Self-blame when outcomes are less than perfect

The neurological basis for this effect has been observed in fMRI studies showing that as decision complexity increases, activity in the anterior cingulate cortex (associated with conflict resolution) and the dorsolateral prefrontal cortex (associated with working memory) becomes increasingly chaotic and inefficient.[14] Eventually, these systems become overwhelmed, leading to what neuroscientists call "cognitive depletion."

When you start talking about all the things your products and services can do and all the possibilities you offer—and, indirectly, all the things that need to change—their mind goes into overload. It can feel like you're trying to change too many things at once. That triggers stress, which triggers cortisol, which triggers "shields up, Mr. Sulu!"

That's why the thrust of Chapter Eight is all about gaining a deep understanding of your customer's "hair on fire" problem, then solving that one problem with value clarity and only that one before moving on to the next. Yes, that runs counterintuitive to what you've probably been taught and practiced. If you have a live one on the line, you want to reel them in with as much as you can, right? But the more new information the human brain has to process, the more likely it will simply shut down, triggering the person's self-preservation orientation.

Remember: Choosing anything new = *risk.*
Choosing multiple options of multiple solutions = *overwhelming risk*

In a high-level sales meeting, you need to be strategic, intentional—surgical, almost—in where you lead the conversation.

Barrier 5: The Scarcity Syndrome Effect

Now, the fifth major barrier to change is that we tend to feel like *we don't have enough resources to implement the needed change.* This is a common one, especially for sales and marketing leaders who feel like they have unrealistic expectations placed upon them and their teams that they simply can't meet. You might feel the same way yourself, like you don't have enough prospects or that you don't have enough tools and systems.

The scarcity syndrome has deep evolutionary roots. For most of human history, resources were genuinely scarce, and conservation was essential for survival. Modern neuroscience has revealed that experiencing scarcity—whether of time, money, or other resources—creates a specific cognitive state that researchers call a "scarcity mindset."[15]

When in this mindset, the brain focuses intensely on addressing immediate scarcity at the expense of long-term thinking or creative problem-solving. Functional MRI studies show that scarcity causes the brain to reallocate neural resources, enhancing the processing of scarcity-related information while inhibiting other cognitive functions.[16]

For sales professionals, this creates a challenging paradox. The very conditions that make your solution valuable—your customer's limited resources—also reduce their cognitive capacity to appreciate and implement that solution.

Now, I'm just going to be real with you: I've found that this is usually an excuse. It's a conscious explanation of a subconscious feeling. Implementing this change or achieving a big goal feels overwhelming. Grug Crood does not take risks. Grug Crood lives in a world of scarcity, fighting for his life every day. Grug Crood lives in a world where he barely survives.

I already have problems with bandwidth. I'm already afraid of losing what I'm risking. I already worry about walking this path alone. Unless you can convince me that I have everything I need to succeed, my default is to assume failure.

The more you talk to me about the million-and-one things your gizmo can do, the more work I hear. My hair's on fire as it is—I don't have time to do all the stuff you're asking!

What kind of support do you offer? Am I going to be on my own (linked to the fear of being isolated from the herd)? What are the contingency plans? Better yet, supply the narrative by telling me about a customer who ran into a challenge and how it was handled.

My caveman brain already perceives a reality of scarcity.

Calm my nerves and prove me wrong. Show me that you have all the resources I could ever need to help create a smooth change process or implementation.

Barrier 6: The Safety Box Effect

The sixth and last barrier we're going to talk about is probably my favorite barrier of all.

When the pressure to change is off, what happens?

We revert back into our safety box.

You see, we've built safety boxes everywhere in our lives. As long as we're in one of our safety boxes, guess how we feel? You guessed it: safe! The minute someone asks us to step out of our safety box, and it's not stepping directly into a different safety box I'm already used to, all of these barriers to change instantly pop up.

We don't want to change. And if we can somehow get rid of the pressure to change, we get to go right back to where we feel comfortable: the status quo. And in a sales conversation, guess where they feel the pressure is coming from? So, if they can get rid of you, they get to go back to the realm of status quo!

The safety box effect is more than just a metaphor—it's a neurological reality. The brain's preference for homeostasis (stability) is managed by the hypothalamus, which works to maintain equilibrium across all bodily systems.[17] Departing from established patterns requires the brain to create new neural pathways, a process that requires significant energy and resources.

This principle is evidenced in what neuroscientists call "functional fixedness"—the tendency to use objects or approaches only in their familiar way.[18] This cognitive bias limits our ability to see alternative solutions or applications, effectively keeping us in our mental safety box.

You see this concept play out every January 1. Everyone's excited about their New Year's resolution. This is going to be the year we finally lose weight and get fit. Thousands of people join the gym. You can't find a free machine and everybody's using all the equipment. But you know human nature. You know that if you're patient enough, by the end of February, the gym will be empty again, and you can go back to your normal routine.

Why?

The pressure to change is off. New Year's is over. The resolution's over. "Meh . . . I'm not really *that* out of shape."

The same thing happens in our customer conversations.

If we don't give our customers a compelling enough reason to change once we leave, guess what they do?

Head right back to their safety box.

Remember: *People don't change until the pain of staying the same is greater than the pain of change.*

You've experienced this dozens of times without realizing it. Your customer will be excited at the prospect of working together, and you shake hands and walk out. The next time you talk to them, you can immediately tell that something's happened. Their excitement has cooled. You hear the worry in their voice. They start going back over issues and potential problems that you thought you'd put to rest.

It's not their fault—it's just human nature. In his book *Misbehaving: The Making of Behavioral Economics*, behavioral economist Richard Thaler wrote, "In physics, an object in a state of rest stays that way unless something happens. People act the same way: they stick with what they have unless there is some good reason to switch."[19] Inertia pulls us right back into our norms. And our preferred norms are familiarity and the desire to be warm and comfortable back in our cozy, albeit somewhat cramped, little safety box.

The Predictable Path of Change Resistance

While these six barriers to change may seem overwhelming, the good news is that resistance to change follows predictable patterns. By understanding these patterns, you can develop strategies to address them proactively.

Researchers Elizabeth Kübler-Ross and David Kessler, known for their work on the stages of grief, observed that people facing significant change often progress through similar emotional stages: denial, anger, bargaining, depression, and acceptance.[20] While originally developed to describe grief responses, this model has proven remarkably applicable to organizational and personal change as well.

This progression explains why initial resistance to your sales proposition doesn't necessarily indicate ultimate rejection. Rather, it may simply reflect your customer's position in their natural emotional journey toward acceptance. The key is to recognize which stage they're in and respond appropriately.

Similarly, change management expert William Bridges identified three phases that people experience during transitions: endings, the neutral zone, and new beginnings.[21] Understanding that your customers must psychologically "end" their relationship with their current solution before they can fully embrace yours provides important context for managing resistance.

Where I used to be perplexed, these days, I recognize that I'm simply dealing with basic neuroscience. I go back to basics and make sure I've addressed the most important questions:

- Have I gained their personal trust and professional respect?
- Have I identified the problem most important to them?
- Have I framed their problem in a way that demonstrates what it's going to cost them if they do nothing?
- Have I shown them an easy way to avoid that loss?

- Have I framed my solution as the best way to solve their problem—not as facts and figures, features and potential benefits, but as the only logical choice to solve the very problem that's preventing them from accomplishing their goals?

- Have I proven that others like them have had success doing the same thing?

- Have I proven that they won't be alone as we implement my solution?

So how do we do that? First, we must understand the psychology of belief and the very biases that may either propel us toward change or drive us farther away.

THE PSYCHOLOGY OF BELIEF AND THE IMPACT OF BIASES

"A man generally has two reasons for doing a thing. One that sounds good, and a real one."

—J.P. MORGAN

THIS CHAPTER IS another foundation-laying chapter. If you'll stick with me, I promise the payoff is worth it in Part II: The Field, when we start bringing all of this together for your next sales conversation.

If we want to understand why people buy, why they resist change, or why they say one thing and do another, we have to go upstream from behavior—to belief. Beliefs are the lens through which we interpret the world, the stories we tell ourselves to make sense of our experiences, and the filters through which all information must pass before it ever reaches conscious thought.

But where do these beliefs come from?

Beliefs are not born—they're built. From early childhood, our brains begin collecting sensory data: what we see, hear, feel, and experience. The brain, in its quest for efficiency and survival, rapidly starts to form connections between these inputs. When something feels repeated, emotional, or significant, our brain stores it not as a one-off event but as a "rule of the world." These rules eventually become beliefs. Whether it's "People like me can't succeed" or "If I do X, I'll be safe," our brains use these beliefs to predict and prepare for future situations.

Most beliefs are formed unconsciously and emotionally first, then rationalized later. Neuroscience shows us that the limbic system—especially areas like the amygdala and hippocampus—encode experiences with emotional weight. The prefrontal cortex shows up later to make sense of what just happened. This means that most of what we believe to be rational is actually rooted in something much older and more emotional.

Once formed, beliefs drive behavior. If I believe I'm not good at public speaking, I avoid raising my hand. If I believe customers are only concerned with price, I'll lead with discounts rather than value. These behaviors become patterns—repeated responses to familiar situations—and patterns repeated often enough become habits.

Here's the catch: the brain loves habits. They're efficient. They reduce cognitive load. And once a behavior becomes habitual, the brain stops questioning whether it's true—it simply accepts it as normal.

This is why belief is so powerful: it creates the behavioral blueprint for our lives. What we believe determines what we do. What we repeatedly do becomes who we are. So when we talk about helping someone change—whether it's a buyer, a team member, or even ourselves—we're not just trying to shift surface-level behaviors. We're challenging the very architecture of belief underneath them.

And that's where biases come in.

Biases are the shortcuts the brain uses to protect our existing beliefs. They're the mental mechanisms that filter, frame, and sometimes distort the way we process new information—especially when that information threatens a belief we've held for years. In the next section, we'll explore the most common cognitive biases, how they show up in buying behavior, and how to navigate them with intention and integrity.

But before we do, remember this: every behavior you see is powered by a belief you don't. If you want to influence behavior, start with belief.

Depending on what research you look at, there are somewhere between 53 and 104 cognitive biases.

"Cognitive bias" is a blanket term that refers to the ways in which the context and framing of information influence our judgment and decision-making.

There are many kinds of cognitive biases that influence individuals differently, but their common characteristic is that—in step with human individuality—they lead to judgment and decision-making that potentially *deviates from rational objectivity.*

In some cases, cognitive biases make our thinking and decision-making faster and more efficient. The reason is that we do not *stop to consider all available information,* as our brain, as the highest calorie-consumptive organ in the body, is looking for efficiency through shortcuts. In other cases, however, cognitive biases can lead to errors for exactly the same reason.

The Adaptive Nature of Cognitive Biases

When we think about cognitive biases, it's easy to consider them as flaws in our thinking—errors to be corrected. However, contemporary neuroscience offers a different perspective. Many cognitive biases actually evolved as adaptive mechanisms that helped our ancestors survive in environments of uncertainty and scarcity.[1]

Gerd Gigerenzer, Director at the Max Planck Institute for Human Development, has pioneered research demonstrating that many biases are actually "fast and frugal heuristics"—mental shortcuts that allow us to make quick decisions with limited information.[2] In ancestral environments where rapid decisions could mean the difference between life and death, these shortcuts were highly adaptive.

For example, what we now call "confirmation bias"—our tendency to seek out information that confirms our existing beliefs—helped our ancestors quickly categorize potential threats without wasting precious cognitive resources. Similarly, "availability bias"—giving greater weight to information that comes easily to mind—helped prioritize recent or emotionally salient dangers.

Understanding the adaptive nature of biases is crucial for sales professionals. Rather than fighting against these deeply ingrained mental processes, effective communicators work with them, using the brain's natural tendencies to facilitate better decision-making.

I could spend all day talking about the psychology of buying, but when you narrow it down to the cognitive biases most meaningful to influential conversations, I believe there are six biases that you really need to have a working knowledge of.

These six biases are driven by the interplay between the limbic system and root brain (the feeling and instinctive brain areas) and the neocortex (the thinking, rational brain). They are:

1. Prospect Theory (Loss Aversion)
2. The Anchoring Effect
3. Confirmation Bias
4. Availability Heuristic
5. Choice-Supportive Bias
6. The Bandwagon Effect

Cognitive Bias 1: Prospect Theory and Why We Hate to Lose

Here's a bizarre but telling game. You have $0 right now. I come up to you and give you one of two choices: I will either give you $50, no questions asked, and you walk away . . . or we can flip a coin: Heads you get $100, tails you get nothing.

Which would you pick?

Again, you have nothing right now. You have absolutely nothing to lose. Sure, you could take the $50, but for really good odds, you could get double that! You have a one-in-two shot of winning the jackpot. When presented with choices like these, people overwhelmingly choose the safe bet of taking the $50 and walking away. Why is that? Why settle for less when there's a 50 percent chance of getting double at zero cost to you?

The answer lies in a book you may have heard of, *Thinking Fast and Slow*. The author, Daniel Kahneman, won a Nobel Prize for his research showcased within it. His research is chock-full of surprising insight. For instance, he found that human beings have two systems of thought or decision-making, the aptly named System 1 and System 2, which we touched on in Chapter Three.

System 1 is our gut, instinctive feelings. We see a person and size them up in a split second. I'm not saying that's a good thing, but it is how our brains work.

System 2 is our more deliberate thought processes. It's us thinking through the order of ingredients we need at the supermarket or working out a math equation in our head.

In NeuroSelling language, System 1 would be a combination of the root and limbic brains and System 2, the neocortex.

But while researching how we make decisions, Kahneman discovered something every salesperson needs to know.

The Prospect Theory: Basically, it says that people don't make decisions based on the final outcome but based on the perception of potential gains and losses. In developing the theory, he stumbled across something sales and marketing people have known—or, at least, suspected—for years.

The Neuroscience of Loss Aversion

What makes loss aversion so powerful? Recent neuroimaging studies have given us a window into exactly what happens in the brain when we face potential losses versus potential gains. When we anticipate losses, there is heightened activity in the amygdala and anterior insula—regions associated with negative emotions and pain processing.[3] In contrast, potential gains activate the brain's reward centers, primarily the nucleus accumbens and ventral striatum, but to a significantly lesser degree than the activation triggered by potential losses.

This asymmetry in neural activation explains why the negative emotions associated with losing $100 typically outweigh the positive emotions associated with gaining the same amount. Remarkably, this neural response occurs even when we intellectually understand that the expected values are identical.

Further research by neuroeconomist Paul Zak has shown that anticipated losses trigger the release of stress hormones like cortisol, which inhibit activity in the prefrontal cortex—the brain region responsible for rationalizing decision-making.[4] This creates a neurological double-whammy: heightened emotional response coupled with reduced rational analysis.

People are twice as motivated to change/take action or remain tied to inaction by potential loss as they are potential gain. It's why you and I would almost certainly take the $50 and walk away happy rather than flipping the coin for a one-in-two chance of $100. You see, our brain has already taken ownership of the $50, so anything that would put that at risk drives our decision-making. The fact that we really had nothing to begin with is a distant memory. Now you're asking me to "risk" $50 for the hope of more. I fear losing my $50 far more than "potentially" gaining $100. My brain becomes fixated on the horrible outcome of being back to zero. This is not logical. It's 100 percent emotional.

So, this idea of avoiding a loss versus pursuing a gain in our sales and marketing message is really, really important. But to be very, very clear, it's not about fearmongering.

You're not trying to go out there and scare your customers to death.

It's about **crafting a narrative that really highlights the cost of the problem** and the **risk associated with not solving the problem** versus how great and wonderful your product is.

The Business Cost of Loss Aversion

The impact of loss aversion extends beyond laboratory experiments into real-world business decisions, often with significant financial consequences. A groundbreaking study by management researchers at the University of Chicago and MIT examined how loss aversion affects corporate decision-making.[5] They found that companies typically require potential gains of two and a half to three times the size of potential losses before approving new initiatives, significantly limiting innovation and growth.

This corporate manifestation of loss aversion helps explain why so many businesses maintain the status quo even when presented with compelling evidence for change. The emotional drive to avoid losses often overwhelms rational analysis of potential gains, creating what economists call "status quo bias" in organizational settings.

For sales professionals, this presents both a challenge and an opportunity. By framing your solution in terms of avoiding losses rather than achieving gains, you align your message with how your customer's brain naturally evaluates decisions. This isn't manipulation—it's communication that works with, rather than against, the brain's inherent decision-making architecture.

All your competition is likely out there with a gain, gain, gain message. (Unless they've already gone through NeuroSelling, in which case you're in big trouble.) "Here's our product, here's how great it is, here's why we're the best, here's why our customers say you should buy our product." That message—and, by association, everyone who says it—becomes white noise. My brain tunes it out.

But when I hear what I'm potentially about to lose, it grabs my attention. The reason?

Because problems evoke emotion. Products evoke judgment. Remember: decisions are made inside out . . . emotionally first, then justified with facts and data.

Talking about what I stand to potentially lose will be processed in my limbic as a "problem." By our very nature, we know unsolved problems can put us at further risk, and that is what the brain wants to avoid at all costs.

That's why, in order to drive urgency to change, you have to frame the problem not as the cost of acting but as the cost of inaction or taking the wrong action.

What do they stand to lose if they do nothing?

Again, I'm not talking about scare tactics. NeuroSelling is not about manipulation. Once you properly quantify the cost of indecision, you won't need to exaggerate or use hyperbole. The numbers themselves should create the urgency. Your job is to make sure they're looking at the right numbers.

Cognitive Bias 2: The Anchoring Effect

At its lowest point, the Mariana Trench is 10,971 meters.

Sounds deep. Can you picture how deep that is? Likely not.

Without any solid point of reference, it's hard for our mind to wrap itself around just how big of a number that is. Even converting it to standard measurements (for us Americans) and learning that it equals 35,994 feet doesn't help much. We need a point of comparison.

Would it help to be reminded that the average, low-end cruising altitude for an airliner is 31,000 feet? Think about looking out your plane window at how far away the ground is. The bottom of the Mariana Trench is 5,000 feet below that. Put another way: if Mount Everest were put in the trench, its peak would still be a mile underwater. Once we have a point of reference, then we can begin to grasp the size and scope of what we're dealing with. We refer to these points of reference as "anchor points."

The Neural Mechanics of Anchoring

The anchoring effect occurs because the brain doesn't evaluate information in absolute terms but rather in relative terms. When we encounter a number or reference point, it activates specific neural patterns that then influence how subsequent information is processed.

Neuroimaging studies have revealed that anchoring activates regions associated with semantic priming—a process where exposure to one stimulus influences the response to a subsequent stimulus.[6] When you present a customer with an initial number (the anchor), it creates a neural framework that influences how all subsequent numbers are evaluated.

What's particularly fascinating is that this effect persists even when people are aware of it and actively try to avoid it. Researchers at Stanford University found that even when participants were warned about the anchoring effect and incentivized to avoid it, they still showed significant anchoring bias in their judgments.[7]

This occurs because anchoring operates largely through the brain's automatic System 1 processes, which are difficult to override with conscious effort. The initial anchor triggers what neuroscientists call a "confirmation bias cascade," where subsequent information processing becomes biased toward confirming the validity of the anchor.

Our same loss-aversion researcher, Kahneman, and a colleague once conducted a study where they asked people to estimate what percentage of African countries were part of the United Nations. But first, the study participants spun a wheel of fortune. The wheel numbers went from zero to one hundred. One group's wheel was rigged to stop at ten, the other's at sixty-five.

Once the participants had spun and gotten their number (ten or sixty-five), the researchers would then ask the participants two questions:

1. Do you believe that the percentage of African countries that are part of the United Nations is higher or lower than the number on the wheel?
2. What do you estimate the actual percentage to be?

The people whose wheel stopped at ten estimated that, on average, 25 percent of African nations were part of the United Nations. The people whose wheel stopped at sixty-five estimated 45 percent.

What did the wheel have to do with international politics and participation in a global body? Nothing.

So why did spinning a random wheel cause one group to nearly double its estimate?

Because of the Anchoring Effect. It works even when there is absolutely zero relationship between two things.

Strategic Anchoring in Business Contexts

The anchoring effect can be observed across virtually all business domains. A landmark study from Harvard Business School and MIT examined pricing decisions in retail, finding that the presence of an original "anchor" price increased sales by 38 percent even when customers knew the discounted price was the standard market rate.[8]

- For sales professionals, this research offers three key strategic insights:
- The first number mentioned in a negotiation often serves as the anchor around which the rest of the discussion revolves

- High-value anchors (e.g., the cost of the problem) should be established before discussing solution pricing
- Non-numerical anchors (conceptual frames, reference points, and stories) can be equally powerful in shaping perception

The retail industry has been using this for years. "Was $299.99, Now $49.99!" Oh, man, I'm saving $250! Was the item really worth $300 originally? Is it even worth $50 now? Who knows?! It doesn't matter. It feels like a great deal because of our brain's point of reference or anchor point.

Retail salespeople use this tactic all the time. You present the customer with a price and let them experience sticker shock. Then you add in the discounts, and voila! A new, better price. Instead of the price looking expensive, it looks like a bargain—I mean, just look at how much it was originally!

As a B2B salesperson, you're not selling cars and furniture, but you're still dealing with basic human psychology. When you're presenting your solution, what's your customer going to compare it to? What's their point of reference?

If you don't give them an anchor, their brain will select it for themselves. Their anchor point might be your competitor they met with yesterday. It could be the article they pulled up on Google five minutes before you walked in. It could be the Wheel of Fortune app they were playing on their phone.

Why leave it to chance? Why not give them a proper anchor point?

Again, nearly 60 percent of B2B sales meetings lead to no action.

Doesn't it seem reasonable that a proper anchor point could be the cost of the status quo? What's it going to cost them to continue doing nothing? What happens if they don't solve the problem they have?

Let me give you an example of how we set the right "anchor" with a sales executive looking to improve their sales effectiveness:

"John, how many salespeople do you have?"

"One hundred."

"What's their individual quota?"

"Roughly two million bucks each."

"What's your goal increase for the year?"

"Jeff," the CEO said, "come hell or high water, we've got to increase top-line revenue by 10 percent this year."

"Well, if I'm not mistaken, 10 percent of $2 million is $200,000, right?"

The CEO says, "Yeah . . . ?"

"So, what you're really telling me is that you have a $200,000 per sales rep problem that has to be solved, correct?"

He frowned for a second, then said, "Well, yeah, I guess we do."

Here, I used the anchor effect.

If I've done my job correctly, his new anchor should have him thinking, "What would I pay per sales rep to solve a $200K problem?" Instead of letting him silently create his own anchor points, I instead helped him anchor the eventual solution I will offer him with how they should have been measuring it in the first place:

What's the cost of not doing anything about this at all or the cost of doing something that won't work?

At $200,000 a rep, anything reasonably south of that begins to look like a bargain!

Now, take a step back and think of it like this: $200,000 increase per rep × 100 reps = $20,000,000.

That's the larger problem that's at risk. Is it safe to assume you could keep doing what you're doing and get halfway there? Probably, but that still leaves you $10,000,000 short! What would you invest to bridge that gap? One percent? Five percent? Ten percent? Likely all of the above, if you felt confident it would get you the results.

This is simply spiking their cortisol—on purpose.

They see what they stand to lose by doing nothing, so they can avoid their fear of loss by going with your solution. I'll show you how to use this strategy later in Part II: The Field.

Cognitive Bias 3: Choice-Supportive Bias

Choice-supportive bias—also called post-purchase rationalization—is the mind's way of confirming that a decision you made was the right one.[9]

In fact, our mind misremembers the actual memory to convince ourselves that we made the right choice.[10]

The Memory Mechanics of Choice Support

The neurobiological basis for choice-supportive bias is fascinating. When we make decisions, the brain doesn't just execute the choice—it actively works to justify it. Research using fMRI has shown that after making a choice, there's increased activity in the dorsolateral prefrontal cortex—a region associated with reappraisal and rationalization.[11]

This process involves what psychologists call "memory reconstruction. Contrary to popular belief, our memories aren't fixed recordings of events but are actively reconstructed each time we recall them. During this reconstruction process, memories are subtly modified to align with our current self-image and beliefs.

Neuroscientist Elizabeth Phelps has demonstrated that this occurs because memory retrieval activates the hippocampus (responsible for memory recall) simultaneously with the amygdala and orbitofrontal cortex (responsible for emotional processing and self-concept).[12] This co-activation allows emotions and self-perception to influence the reconstructed memory.

For sales professionals, this explains why challenging a customer's previous purchase decisions often backfires. When you suggest that their current solution is inadequate, you're not just critiquing a product—you're contradicting the rationalized memory that justifies their previous choice.

This is partially what I came up against speaking to those surgeons at the conference in Denver. They'd originally made the choice to use the other surgical method. It didn't matter to them that the original reasons (it was the best at the time) were no longer valid. Their unconscious choice-supportive bias prevented them—or at least hindered them—from objectively examining new alternatives.

In studying choice-supportive bias, one study found that company board members who voted for hiring the CEO were more likely to overlook their faults and stick with them through challenging times than board members who weren't.[13]

That is, just the fact they made a choice and were "invested" in a CEO automatically made them give the CEO extra slack—slack they wouldn't give if they'd become board members the day after the vote. In other words, at the subconscious level, if the CEO failed, so did they.

Studies have shown that voters who weren't certain about which politician to vote for before going into the polling booth showed increased confidence coming out. Just the very fact that they'd made a choice increased their confidence that they'd made the right choice.

Strategic Implications for Sales

The implications of choice-supportive bias extend far beyond post-purchase rationalization. It affects the entire customer relationship and creates significant challenges for competitors trying to displace an incumbent provider.

Research from Stanford University examining B2B purchasing decisions found that companies typically require two and a half times more evidence to switch from an existing provider than they would need to make an initial purchase decision.[14] This "incumbent advantage" explains why even clearly superior products often struggle to displace established solutions.

What's particularly interesting is that the strength of choice-supportive bias is proportional to the difficulty and importance of the original decision. The more significant the original purchase decision, the more powerful the subsequent bias becomes. This explains why high-stakes B2B purchases (like enterprise software or medical equipment) create especially strong psychological barriers to switching.

With your customers, be hyperaware of their choice-supportive biases. Whatever choices they've made in the past make sense to them (even if their brain has to make up some of those reasons since then). It doesn't matter that the whole reason they made the decision in the first place doesn't exist anymore. They'll create new reasons to justify their past choices. In other words, their brains say, "You made this choice. Therefore, it must be the right choice."

That's why starting with facts and figures, features, and benefits (selling from the outside in) doesn't work. The mind can justify why a person's original choice was superior, even if that means rewriting the memory. **You're not selling based on facts but on feelings—especially the ones your customer is unaware of**.

The key is, I can't feel that you're telling me the choices I've made to this point were wrong.

I have to arrive at that conclusion and the conclusion to choose you all on my own. Once I do, I'm yours forever. Or until you don't live up to your promise. Which I'm sure will never happen.

Cognitive Bias 4: Confirmation Bias

Whatever your political stance, you almost certainly see news articles and online posts that confirm things you already know to be true. All the time. The danger is that **our minds give extra weight to the things we already believe and easily dismiss or explain away evidence to the contrary**.

The Neurobiology of Belief Persistence

Confirmation bias is one of the most thoroughly studied phenomena in cognitive psychology, and its neurological underpinnings reveal why it's so persistent and pervasive. When we encounter information that confirms our existing beliefs, the brain's reward centers—particularly the ventral striatum—show increased activity, delivering a small "neurological reward" for finding confirming evidence.[15]

In contrast, when we encounter contradictory information, the brain's conflict-detection mechanisms in the anterior cingulate cortex activate, creating a mild state of cognitive discomfort. This discomfort triggers defensive processing in the dorsolateral prefrontal cortex, which works to discredit or minimize the contradictory information.

What's particularly striking is that this process happens automatically, before conscious thought. Neuroimaging studies have shown that the brain begins processing information differently based on whether it confirms or contradicts existing beliefs within the first 200 milliseconds—far too fast for conscious evaluation.[16]

This automatic filtering creates what psychologists call an "asymmetric scrutiny effect," where confirming information is processed fluently while contradicting information undergoes rigorous (and often biased) scrutiny.

If I believe Chevys are the best pickup truck on the market, I'm always going to find magazine articles and hear people's stories about how great they are. If I see a news story about a navigation error resulting in a wreck, I'll chalk it up to a fluke or a bad batch.

The human mind will go to incredible lengths to cling to its beliefs, using the neocortex to rationalize completely irrational behavior.

In *The Psychology of Influence*, psychologist Robert Cialdini recounts the story of the cult who believed space aliens were coming back for them at midnight on a certain date. Midnight came and went. Scotty didn't beam anybody up. Some of the cult members came to their senses and realized they were in crazy town, so they left. But most of them stayed. Around 4:00 a.m., one of their leaders begins receiving messages that the aliens had changed their mind because the cult members had done such a good job getting everyone ready. Instead of facing the reality that someone was either scamming them or was outright delusional, the cult members renewed their zealotry and became even more engaged with the cult. In the absence of reason, their minds *invented* reason.

The "Backfire Effect" in Sales Contexts

One of the most troubling manifestations of confirmation bias in sales contexts is what psychologists call the "backfire effect"—where presenting contradictory evidence actually strengthens a person's original belief.[17] When strongly held beliefs are challenged, the defensive response can actually reinforce the very belief you're trying to change.

This explains why traditional "feature-benefit" selling often fails, especially when trying to displace an existing solution. Each competitive claim you make can actually strengthen your customer's commitment to their current provider through this backfire mechanism.

A study at Dartmouth College examined what happens in the brain when political partisans were presented with contradictory information about their preferred candidate. Rather than changing their minds, the contradictory information activated reward centers in the brain as participants found ways to maintain their original beliefs.[18] The more contradictory information presented, the more rewarding it became to defend their position.

For sales professionals, this creates a paradox: the more aggressively you present evidence against a competitor, the more likely you are to strengthen your customer's loyalty to that competitor.

There are a thousand heartwarming stories about people overcoming racial biases and ethnic stereotypes. Just about every time, it started with two people finding common ground, then finding connection, then finding friendship.

This is why your Why Story is so important to a successful conversation.

By finding common ground, sharing some universal beliefs, and then creating a connection, you've bypassed the potential confirmation biases the neocortex had at the ready. You've made "backdoor friends" with the heartwarming emotional brain.

Cognitive Bias 5: Availability Heuristic

I've learned the hard way that when I think about something, the first thing that comes to mind is usually incorrect. I have mostly life and my wife to thank for this.

My natural tendency is to fall prey to my availability-heuristic bias. Basically, my brain assigns more importance, weight, and validity to information I can easily recall or have immediate access to than it does to external information—especially if that information contradicts what I believe I know. The result is that, unless I'm careful, I'll let my fears and beliefs override good sense.

The Evolutionary Logic of Availability

The availability heuristic evolved for good reason. In ancestral environments, events that could be easily recalled were often those that occurred frequently or recently—making them reliable guides for decision-making. If you could easily recall instances of a certain plant making people sick, avoiding that plant was an adaptive response.

Neuroscientific research has shed light on why availability is such a powerful influence on judgment. When we retrieve easily accessible memories, the process activates the hippocampus (memory retrieval) simultaneously with the amygdala (emotional processing) and anterior cingulate cortex (evaluation).[19] This creates a powerful feedback loop where easily recalled examples carry outsized emotional weight in decision-making.

What makes this particularly relevant for sales professionals is that the availability heuristic affects even highly analytical individuals making significant business decisions. A study of investment bankers found that their risk assessments were heavily influenced by the availability of recent examples, even when those examples were statistical outliers.[20]

You've probably met someone who doesn't "believe" in seat belts or that smoking can kill them. They'll say things like, "I know a guy who lived through a car crash. They say if he'd been wearing his seat belt, he'd be dead. That's why I never wear the things!" "My grandfather smoked three packs a day all his life and lived to be ninety. You can't believe everything you read." They've extrapolated a dangerous set of beliefs from a very limited amount of information.

How does this apply to sales?

Say you were a tech vendor selling enterprise software. As part of your standard lead-in, you refer to how fast your business is growing in the industry and why the customer should invest in your cutting-edge SaaS. That morning on the drive into work, however, your customer heard that the tech-heavy NASDAQ had its biggest fall of the year yesterday.

It doesn't matter that it ended higher than at any point in the first quarter. It doesn't matter that the overall trend of the NASDAQ is positive. The customer's availability heuristic gives more weight to what they can immediately recall.

Arguing with them by presenting facts and economic trends isn't going to change their mind. In fact, they'll probably just become more entrenched in their stance that things look bad for tech—they'd be better off just sitting tight until they see which way the wind's blowing.

Cognitive Bias 6: The Bandwagon Effect

Of the cognitive biases we've discussed, this is probably the easiest and most familiar. As human beings, we're wired to be part of the tribe. **We don't like going it alone or trying something first**. (Recall the change barrier of feeling isolated.)

Remember the old saying in the eighties, "Nobody ever got fired for choosing IBM?" Big Blue had that kind of a reputation in enterprise-level business. Everybody knew about IBM. Everybody.

The Social Brain and Collective Decisions

The bandwagon effect stems from what neuroscientists call our "social brain"—the neural networks dedicated to understanding and navigating social relationships. This system includes regions like the medial prefrontal cortex, temporoparietal junction, and superior temporal sulcus—all of which help us monitor social norms and align our behavior with the group.[21]

When we observe others making certain choices, these brain regions become activated in ways that predispose us toward making the same choices. This occurs through two distinct neural mechanisms:

1. Social proof processing – When we see others choosing a particular option, the brain's ventral striatum (associated with reward anticipation) shows increased activity, making that option seem more valuable[22]

2. Social risk evaluation – The anterior insula becomes less active when choosing options that others have chosen, indicating a reduced perception of risk[23]

These neurological responses aren't conscious or deliberate—they're automatic processes that evolved to help humans navigate complex social environments. They explain why B2B decision-makers consistently report that peer recommendations and case studies from similar companies are the most influential factors in their purchasing decisions.

Have you ever stopped to wonder how much money IBM made just from people feeling that way? IBM was seen as the safe choice. Sure, there were a few other games in town, but when people are under pressure, they choose safety. In this case, safety in numbers, aka "the bandwagon effect."

In sales, I've seen reps spend so much time talking about options and potential benefits, yet they never use one of the most powerful cognitive biases—giving the customer a chance to get on the bandwagon! As we discussed earlier, there's safety in numbers. There's less risk when we move as a group than in being the Lone Ranger.

We're wired to need that social proof and validation.

Your customer needs to know that they're not hopping on this bandwagon alone. They need to know that others have tried this and found it safe. The more people who do something—our minds reason—the less dangerous it is. The safer they feel, the lower their cortisol. The lower their cortisol, the more they can engage with what you're saying and the more they see others are on board, the more FOMO or "fear of missing out" starts to work on your behalf.

From Biases to Business: The Applied Neuroscience of NeuroSelling

Understanding these six cognitive biases is more than an academic exercise—it provides the foundation for a radically different approach to sales conversations. When you align your communication strategy with how the brain naturally processes information and makes decisions, you create what neuroscientists call "processing fluency"—the ease with which information moves through the brain's evaluation systems.[24]

Research has consistently shown that when information is processed fluently, it is:

- Perceived as more truthful
- Judged as more valuable
- More likely to influence decision-making
- Better retained in memory
- More likely to be shared with others

By incorporating cognitive biases into your sales approach, you're not manipulating decisions—you're facilitating them. You're removing the neurological friction that often prevents optimal decision-making.

This "friction reduction" approach aligns perfectly with how customers want to buy. A landmark study by Gartner found that B2B customers now spend only 17 percent of their purchase journey meeting with potential suppliers.[25] The rest of their time is spent researching independently or processing the information they've received. By understanding and working with cognitive biases, you can ensure your message continues to work for you even when you're not in the room.

Back in Chapter Five, we covered the brain's barriers to change. Here in Chapter Six, we've covered biases and how to neutralize them or even get them to work for you. But the best tool you have at your "change" disposal is actually your customer's amazing storytelling brain.

Let's talk about the surprising neuroscience of narratives.

THE NEUROSCIENCE OF NARRATIVE

"Never tell a story without a point and try not to make a point without a story."

—Papaw Willie Bloomfield

TOWARD THE BEGINNING of our workshops, I always share my Papaw story. I know the old truism is true: *"People don't care how much you know until they know how much you care."*

Unless the people listening to me trust me as a person, I know their shields are going to stay up. Their neocortex is going to critically judge everything that comes out of my mouth. Sharing my Why Story helps them see me as a relatable human being, not just another talking head.

At some point later in the first day, I'll give them a rapid-fire pop quiz. (Do the exercise yourself: Pause for a second after each question and see if you remember the answers too!)

"How many acres was Papaw Bloomfield's farm?"

"A hundred acres!" everybody shouts out.

I don't even give them a moment to pause before the next question: "How long was his driveway?"

"Fifty yards!"

"How much education did he have?"

"Eighth grade!"

"What kind of tractor did he have?"

"John Deere!

"What was the color and make of his pickup truck?"

"Green Chevy Silverado!"

I'll stop and let them catch their breath, then I'll ask, "Now, how in the world did you remember all that information collectively, in unison?"

I love the slightly confused or even surprised look on their faces as they ask themselves the same question: *How DID I remember all that*?!

If they closed their eyes, they probably couldn't tell you what color shirt I'd been wearing all morning, yet they can recall details from a second-hand account told hours and, sometimes, days before.

It's not some kind of mass hypnosis or voodoo. I just know how to use neuroscience.

Stories and the Brain: A Revolution in Understanding

In the past decade, advances in neuroimaging technology have allowed scientists to observe what happens in the brain during storytelling with unprecedented clarity. These findings have revolutionized our understanding of how narratives affect human cognition, emotion, and decision-making.

When we process factual information like statistics or product features, only the language-processing regions of the brain (primarily Broca's area and Wernicke's area) show significant activity.[1] However, when we engage with a story, something remarkable happens—multiple regions across the brain activate simultaneously.

Neuroscientist Uri Hasson at Princeton University pioneered research demonstrating that during effective storytelling, the brain doesn't just process language; it creates immersive simulations of the narrative events. Using functional MRI, Hasson's team discovered that stories activate the same brain regions that would activate if we were actually experiencing the events being described.[2] For example, when you hear about someone running, your motor cortex becomes active. When you hear about delicious food, your sensory cortex lights up.

This "neural coupling" effect explains why stories create more vivid memories and deeper emotional connections than facts alone. Your brain isn't just receiving information—it's living it.

Narratives, Internal Visualization, and Recall

There have been lots of studies on information retention and recall. The way you communicate information will determine how it's received and ultimately stored in the customer's brain. How does that work? What I'm about to share probably won't surprise you since you now have a rough idea of how the brain is wired.

Studies by the World Bank and National Training Laboratories, as well as others, show consistently that 5 to 10 percent is about all you're going to retain of facts, data, and such information.[3] The London Business School, as well as Stanford University, did similar, fascinating studies on retention, though, teaching different groups of students the same information but in different ways.[4]

The first group just received the facts and data. The second group was presented with the same facts and the data, but they threw in a couple of visuals. With the last group, they communicated the same exact information but through storytelling.

The group that got just the facts and data retained about 5 to 10 percent. Not surprising. The group that also had a few visuals retained about 20 to 25 percent of the information. The group that received the information via storytelling retained between 65 and 70 percent—about triple!

The Neurochemistry of Storytelling

What makes stories so much more engaging and memorable? The answer lies in their unique ability to trigger powerful neurochemical responses. When we experience an effective story, our brains release a cascade of chemicals that enhance attention, emotion, and memory formation:

Oxytocin: Often called the "trust hormone," oxytocin is released during emotional narratives, particularly those involving character struggles. Research by neuroeconomist Paul Zak has shown that character-driven stories consistently trigger oxytocin synthesis, increasing feelings of empathy and connection.[5] This explains why personal stories like my Papaw narrative create such deep trust.

Dopamine: When a story creates suspense or anticipation (like wondering what will happen next in a narrative), the brain releases dopamine, which aids in focus and memory formation. Neurobiologist John Medina's research demonstrates that dopamine release makes information up to 20 times more memorable.[6]

Cortisol: Stories involving tension or conflict trigger cortisol release, which helps focus attention and create a sense of urgency—precisely the emotional state that makes information seem important enough to remember.

Endorphins: The emotional resolution at the end of a well-crafted story triggers endorphins, creating positive associations with the message and its messenger.

This neurochemical cocktail creates what scientists call an "emotionally enhanced memory trace"—a memory that's both more vivid and more accessible than those formed through purely factual learning.

Remember: *Narratives engage our limbic and root brain.*

We have an internal visualization mechanism that recreates the experience as if it were our own.

How does this internal visualization mechanism work?

I want you to read until the end of this paragraph, then close your eyes for a moment to "see" this in your mind's eye. Pretend you're standing on the street where you grew up. You're looking back at the house, the trailer, the condo, the apartment, or whatever you grew up in.

Can you see it? Do you remember it? Was it brick? Was it wood? Was there siding? Was the door red? Was it black? Was there an oak tree in the front yard? What did it look like?

You can start to feel your senses come alive. Could you hear another car driving by? Did you feel the wind blow? Do you remember any special smells from your neighborhood?

Right now, you're looking at the words of this book. That image travels down your optic nerve and gets reflected back on the occipital lobe of your neocortex. That's how you literally see with your eyes. When you internally visualize something, those recalled images travel via the hypothalamus region of the limbic system and get projected onto your imagination.

When I can get you to see something using your limbic system, you become completely engaged.

That information gets stored differently than facts in your neocortex. "Experienced" memories—whether real or not—get stored in long-term limbic memory.

Neural Coupling: When Two Brains Synchronize

One of the most fascinating discoveries in narrative neuroscience is the phenomenon of "neural coupling" or "brain-to-brain synchronization." Princeton University did a study where they showed that when the communicator uses a story-based or a visual storytelling approach to communicating, not only does the brain of the listener light up in the right spot, but it also actually lights up in the exact same spot that the communicator's brain is lighting up as they're telling the story—just like the monkey watching the other monkey eat ice cream.[7]

This synchronization and harmonization of two brains when you're communicating through storytelling is exactly how the brain is wired to receive that information, assigning emotion to that information to help drive decision-making.

When we tell stories, our neural patterns can actually synchronize with those of our listeners, creating a remarkable mind-meld effect. Using sophisticated neuroimaging techniques, researchers led by Uri Hasson demonstrated that during effective storytelling, the same brain regions activate in both the storyteller and the listener, often with a slight delay.[8]

This neural coupling doesn't happen during the transmission of pure data or unrelatable information. It specifically occurs during narrative communication that contains emotions, sensory details, and relatable experiences. The stronger the neural coupling, the greater the understanding, empathy, and retention of the information.

This biological synchronization explains why human beings have used stories as our primary mode of communication throughout history—long before written language existed. Stories don't just deliver information; they create shared experience at the neural level.

So far, we've been talking about internal visualization: relying on my mind's eye to create an image of the picture you're painting with words. But what if you took it a step further?

What if, instead of having them conjure up the images, you created the visuals?

I want to challenge you to think a little differently about the visual aids you may be using. Most sales and marketing professionals try to use facts and data as their visual aids (à la PowerPoint). As a sales professional, think about your sales deck right now. How many slides is it? How many bullet points are on each slide?

Just know this: a bullet point is typically interpreted in the customer's mind as a nonemotional, lifeless data point. If that's how you're communicating, even on slides, you're communicating to the neocortex.

Story vs. Features: The Neuroscience of Buying Decisions

The distinction between how the brain processes stories versus features has profound implications for sales professionals. Neuroscientists at Stanford University conducted a study comparing brain activity when people were presented with product features versus a story about how the product solved a real problem.[9]

When participants encountered feature lists, their brains showed activation primarily in the Wernicke's and Broca's areas—regions associated with language processing. However, when they engaged with stories about the product in use, their brains lit up in regions associated with experience, including the sensory cortex, motor cortex, and limbic areas associated with emotion.

This research explains a phenomenon many sales professionals have observed: customers who can visualize themselves using and benefiting from a product are far more likely to purchase it than those who simply understand its specifications.

A B2B Narrative Example

As you can imagine, I've had plenty of business professionals hesitant to use narratives. It's basic neuroscience: I'm telling them that they should try something new. It's different than what they've been doing. It's outside of their safety box.

Of course, your inner caveman Grug Crood is going to say no.

They'll say that their customers are sophisticated or that they don't respond to cute little stories. They might say that they don't need or want to dumb down their information. But what they're really saying is, "I'm not comfortable with this."

They don't realize that they're not only working against basic neuroscience, but they're also doing a disservice to their customers. Our brains can only process a finite amount of information in our short-term memory. Remember back in school when the teacher was presenting a brand-new idea? After listening to her lecture for an hour, your brain would feel overloaded. Your eyes would glaze over, and drool would start coming out of the corner of your mouth. She might as well not even have bothered with the last ten minutes—you had mentally checked out waaayyy before that.

Overcoming Cognitive Limitations Through Story

This limited processing capacity has been well-documented in cognitive science. Research by Princeton University psychologist George Miller established that working memory can only hold approximately seven plus or minus two pieces of information at once.[10] When we exceed this capacity—as most product presentations do—information simply fails to register.

Stories help overcome this limitation through a process called "chunking," where multiple pieces of information are organized into a single meaningful unit. When information is embedded in a narrative, it becomes part of a coherent whole rather than isolated facts competing for limited cognitive resources.

Neuroscientist Daniel Levitin explains that stories provide a "cognitive container" that helps the brain organize and process information more efficiently, significantly expanding our effective mental bandwidth.[11] This explains why stories can convey complex concepts that would be overwhelming if presented as disconnected facts.

Our brains have higher recall when new memories are created in connection with existing neural connections. Here's an example: Let's say you're selling a brand new, state-of-the-art piece of software that does everything I could ever want in my business. From accounting to payroll to project management and CRM, it's an all-in-one solution.

You could start our conversation by telling me all the stuff it can do. I'll feel like I'm standing in front of a fire hydrant, but at least you'll have done your job, right? I tuned out three minutes into your spiel, but sure, tell me everything you possibly know about your product.

What if, instead, you took out your iPhone and said,

> *Jeff, I remember when phones just made calls and sent text messages. I used to carry around a calculator, a laptop, and a digital camera. Shoot, I remember sitting in McDonald's parking lots with a Rand McNally map, trying to figure out where I'd made a wrong turn and worrying that I'd be late for my next sales meeting. I thought getting a turn-by-turn Garmin was pretty high-tech.*

> *I can't imagine going back to those days. I have an app or three for each of those functions. My phone has a better camera than the digital one I used to carry around and I don't need to constantly take out the SD card and import my pics into my computer. I use my iPhone to do everything. In fact, more than I ever thought possible with just a phone. And with everything in the cloud, I don't have to worry about losing my phone or it getting stolen. I just pick up a new phone, log into the cloud, and I'm right where I left off.*

In my head, I'm silently agreeing. That's been my experience too. I remember when gas stations used to sell maps right by the checkout and how frustrating it was when I went to take a picture only to see MEMORY CARD IS FULL flash across the screen.

Now, Jeff, that's how our SaaS platform works. You don't need one system for accounting and another for CRM, then outsource your payroll to a third-party. Nor do you need to worry about having a dedicated IT guy taking care of the server in your closet. Just like your iPhone, you have all the apps to run your business in one place, and you don't ever have to worry about hardware. It's all in the cloud. Our product works just like an iPhone.

Instead of presenting me with new information, you activated information that I already believed to be true, then piggybacked on my existing knowledge to make a link with something else.

Grug Crood doesn't automatically smash it with a club because it looks like something that my brain has already accepted.

When your mind makes the connection with an experience or concept you already know, it forges a neural link between those two isolated stories. **Not only do you understand the other person better, but you'll also remember what they're saying better as well.**

The Neuroscience of Analogies and Metaphors

The iPhone story works because it employs what neuroscientists call "neural reuse"—the brain's ability to repurpose existing neural circuits for new functions.[12] When we understand a new concept through analogy or metaphor, we're literally repurposing existing neural pathways to process new information.

This mechanism is remarkably efficient. Research using electroencephalography (EEG) has shown that metaphors and analogies significantly reduce the cognitive load required to process new information.[13] When information is presented through a familiar metaphor, comprehension speed increases by up to 40 percent compared to literal descriptions of the same concept.

The power of analogies extends beyond just comprehension. Studies have shown that concepts learned through analogies are often:

- Recalled more accurately
- Applied more appropriately to novel situations
- More resistant to forgetting over time
- More likely to influence decision-making

This explains why skilled communicators frequently use analogies to explain complex products or services. By mapping new information onto familiar frameworks, they significantly reduce the cognitive effort required for understanding.

If I go into a sales conversation and dump a load of random facts and figures, you're asking my brain to process and store all this stuff separately. The minute you make a connection between your new information and something I already know, my brain creates a whole load of neural connections. You'll literally engage more of my brain.

The Two Ways the Brain Stores Information

Quick question: how many outside doors did the house you grew up in have?

Did you immediately and automatically see your house in your mind's eye? Even if just for a few seconds, did a picture reel play in your head as you quickly went through and around the house, counting the outside doors?

Those are memories your brain has stored visually.

Next question: could you explain mirror neurons to me?

Your mind probably didn't reach for a picture—you probably started thinking back to the words you read a few pages ago. We store abstract concepts like that verbally.

Not all memories are created equal.

You see, the brain stores memories in basically two ways: verbally and visually.

Our visual recall is far stronger than our verbal recall.

The Dual-Coding Theory of Memory

This distinction between visual and verbal memory aligns with what cognitive psychologists call "dual-coding theory," first proposed by Allan Paivio.[14] According to this theory, the brain encodes information through two distinct but interconnected systems:

- The verbal system: Processes and stores linguistic information as words, concepts, and propositions

- The visual system: Processes and stores information as mental images, spatial relationships, and sensory impressions

When information activates only one system (typically the verbal one), recall is limited. However, when information activates both systems simultaneously—as stories with vivid imagery do—memory formation is significantly enhanced.

Neuroimaging studies have confirmed this theory, showing that when both systems are engaged, there is increased activity in the hippocampus—the brain's memory formation center—suggesting stronger memory encoding.[15]

This dual processing explains why facts embedded in stories are remembered so much better than facts alone. A compelling story activates both your verbal and visual neural networks, creating multiple pathways to the same information and dramatically improving recall.

It's why we can't remember the words to the Bill of Rights we learned in our high school civics class and yet know exactly what clothes we were wearing that time we seriously embarrassed ourselves in that same classroom.

When you store information both visually and verbally, however, you have exponentially higher recall.

That's why facts embedded in stories are so much more powerful than facts alone. Jerome Bruner, the godfather of clinical psychology, found that when facts are embedded in narratives, people are twenty-two times more likely to recall them.[16]

If, at the start of a workshop, I verbally listed those random facts from my story—a hundred acres, a fifty-yard driveway, an eighth-grade education, and a green Chevy pickup—almost no one would recall those details hours later. But because those facts were stored both visually (in their mind's eye as they "experienced" that story with me—ah, the power of mirror neurons!), as well as verbally, virtually everyone I try this with has near-perfect recall . . . on a second-hand story they heard from a total stranger.

Wouldn't it be great if your customers had near-perfect recall of everything you told them?

Now that you know the science behind how we store memories, achieving this is pretty straightforward: simply embed those facts, figures, and data in a narrative.

The Business Impact of Narrative-Based Selling

The neurological advantages of narrative-based selling translate directly into business results. When examining the effectiveness of different sales approaches, we have found consistent patterns:

1. Increased conversion rates: Sales presentations using narrative frameworks show consistently higher conversion rates compared to feature-based presentations.

2. Larger deal sizes: When customers engage with stories about how products solve problems, average deal sizes tend to be higher

3. Shortened sales cycles: Using narrative approaches reduces the average B2B sales cycle, primarily by reducing the "consideration" phase.

4. Improved cross-selling: Customers who received narrative-based explanations of solutions were more likely to purchase additional products or services.

5. Higher customer loyalty: Retention rates for customers acquired through narrative-based selling appear to be higher than those acquired through conventional approaches.

Not surprisingly, when examining brain activity during purchasing decisions, customers who are exposed to narrative selling show significantly different neural activation patterns than those who receive traditional feature/benefit presentations. The narrative group shows stronger activity in regions associated with trust, reduced perception of risk, and positive emotional associations—all critical factors in buying decisions.

Let's go a little deeper into crafting effective narratives.

The Basic Elements of an Effective Narrative

There is a power in story, and it doesn't have to be the "once upon a time" variety. Great sales and marketing professional communicators understand that evoking a visual narrative is simply about using tried-and-true techniques to convey a message in different ways.

Some people are born storytellers. Papaw Bloomfield was one of them.

The great thing about NeuroSelling is that you don't have to be one of these people. You don't need to master the art of storytelling. Creating a compelling narrative isn't that hard because any story is better than no story. **You just need the basic elements and an elementary structure.**

The Hero's Journey in Business Storytelling

While simple stories can be effective, the most powerful business narratives often follow what mythologist Joseph Campbell identified as "The Hero's Journey"—a universal story structure found across cultures and throughout history.[17] Modern neuroscience has revealed why this particular narrative structure is so effective.

When we encounter a hero's journey narrative, it activates our emotional/empathic network—a set of interconnected brain regions associated with self-reflection, empathy, and future planning. This network helps us project ourselves into the story and imagine how we would respond in similar circumstances.

The classic hero's journey includes several key stages that map remarkably well to the customer buying process:

- The Ordinary World – The current state (status quo)
- The Call to Adventure – Recognition of a problem or opportunity
- Refusal of the Call – Initial resistance to change
- Meeting the Mentor – Finding a guide (your solution)
- Crossing the Threshold – Making the decision to change
- Tests, Allies, and Enemies – Implementation challenges
- The Reward – Achieving initial results
- The Road Back – Integrating the solution
- The Resurrection – Transformative results
- Return with the Elixir – Becoming an advocate

This framework works because it mirrors the psychological and emotional journey that all humans experience when facing change—a process that's deeply encoded in our neural architecture. By mapping your customer's buying journey onto this familiar structure, you create what psychologists call "narrative transportation"—a state where the listener becomes immersed in the story and less resistant to its message.

The easiest and most traditional approach is the straightforward model we've used for centuries.

First, you need the setting. Then you need a **good guy** (main character/protagonist). The good guy has goals, hopes and dreams. Then, along comes the **bad guy** (antagonist). The bad guy attempts to prevent the good guy from achieving his hopes and dreams. But alas, along comes **a sage**, a guide who shows the good guy a way forward. A way to overcome the bad guy. The good guy has to make a choice and take action. Then comes the **outcome** (did they succeed or fail?) and the **moral**.

From the tortoise and the hare to Cinderella to Odysseus, they all follow this basic structure. This structure follows the tried and true Joseph Campbell "hero" story structure.

Sometimes, the bad guy isn't an enemy but a situation. For instance, picture two mice in a maze. Every day, they get up, race through the maze to find the big block of cheese, and then bring some back to their cage. One day, the smarter mouse notices the block of cheese has gotten smaller over time. He decides to start spending part of each day searching out the rest of the maze for a new block of cheese. The other mouse thinks it's a waste of time.

The smarter mouse finds some more cheese and starts memorizing the route to get there. Eventually, the original block of cheese is gone. The smarter mouse is fine because he's already been nibbling off the new block. The other mouse starves to death.

This simple business fable sold millions of copies of *Who Moved My Cheese?* and gave people an easy metaphor for a complex message: "We need to be innovating and finding new markets because one day our current market will be disrupted."

I've helped sales teams create hypothetical stories around everything from mountain climbing, preparing for tornadoes, and a floor supervisor in charge of a factory line. You don't need a riveting Shakespearian drama.

You just need a structured narrative.

Character Roles in Sales Narratives

The most common mistake salespeople make is positioning themselves or their company as the hero of the story. This triggers what psychologists call "reactance"—a form of psychological resistance that occurs when people feel their choices are being restricted.[18]

Instead, effective sales narratives follow a specific character structure:

- The Customer is the Hero – The protagonist of the story facing challenges
- The Problem is the Villain – The antagonist preventing success
- You are the Guide/Mentor – The sage who provides wisdom and tools
- Your Solution is the Magic Weapon – The special tool that helps defeat the villain (problem)

This structure works because it aligns with how our brains naturally process narrative information. Studies using fMRI have shown that when we position the customer as the hero, their brain's reward centers activate as they imagine their own success, creating positive associations with the solution being presented.[19]

In Chapter Nine, we're going to talk about the "7 P's": the five narratives you need when walking into a sales conversation. Once you have this framework, you'll be able to create trust-building, change-inducing narratives for any given customer conversation. Now, the fun begins. Let's take what we've learned in the lab and apply it in the field.

From Science to Strategy: The Neural Advantage

As we transition from the science of narrative to its practical application in sales conversations, it's worth summarizing the key neurobiological advantages that stories provide:

1. Enhanced memory encoding through dual-coding of information (verbal and visual)
2. Increased neural engagement across multiple brain regions
3. Positive neurochemical responses (oxytocin, dopamine, endorphins)
4. Neural coupling between speaker and listener
5. Reduced cognitive load through chunking and pattern recognition
6. Bypassed resistance by engaging the limbic system before the critical neocortex
7. Increased persuasive impact through emotional engagement

By understanding these mechanisms, you can craft sales narratives that work with—rather than against—the brain's natural information processing systems. In the chapters that follow, we'll translate this neurological understanding into practical frameworks and strategies for your customer conversations.

To Learn More About NeuroSelling®, go to
www.braintrustgrowth.com/neuroselling

PART II

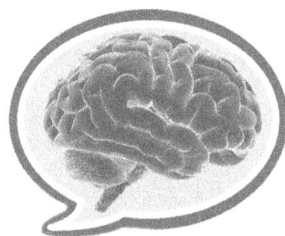

THE FIELD

8

START WITH WHO, LEAD WITH WHY

"Customer-centric . . . means you give the customers what they want rather than what you want to sell them."
—JACK MITCHELL, HUG YOUR CUSTOMERS

WE'VE SPENT THE first seven chapters exploring the scientific foundations of how the human brain makes decisions. Now it's time to transform that knowledge into practical strategies that will revolutionize your customer conversations. As we transition from theory to application, remember the key principle we've established: the brain processes information from the inside out—emotion first, logic second.

This chapter introduces the first step in the NeuroSelling methodology: understanding your customer deeply before attempting to influence them. This approach aligns perfectly with how the brain naturally makes decisions and forms connections.

From Theory to Practice: Applying the Neuroscience of Influence

Once upon a time, I coached individual executives, not sales teams.

It seemed that every time I sat down with a president or CEO, the conversation would go, "My biggest problem right now is we're not growing. If you can coach me through how to fix that problem, that would be huge."

In my head, I was thinking *nearly every growth problem can be tied back to leadership—that's the whole reason we're doing this coaching! Let me help you*! But telling a CEO that his whole problem is between his ears isn't a particularly popular message.

Nor is it an effective way to help people. You see, because I understood how the brain works, I knew that the CEO's self-preservation orientation was triggered. His worry was, "If I don't right this ship, the board's going to throw me overboard. How do I save myself?!" Until we could solve that problem, his brain was hyperfocused on the imminent threat. As long as he felt like he was in the trenches under fire, there was no way I could pull him into the general's tent and talk long-term leadership development.

With these clients, I finally realized that before I could earn the right to speak into their personal development, I needed to help each of them address the overwhelming problem in their professional life.

"Alright, sometimes it helps to have fresh eyes on the situation or just an objective third party in the room. Let's get your VP of sales in here, maybe your head of marketing, and see if we can do something different."

My customer couldn't hear what I had to say over the deafening noise of their immediate fear/problem. **They couldn't focus on what really *needed* to be fixed until they were certain we corrected what *had* to be fixed. Sales.**

The broader point here is a question that must be asked. How well do you really know your customers? How well do you know their role, their top three goals/objectives, and the impact their success or failure has on their department or their company? What are the top problems that prevent them from accomplishing their goals/objectives? If I asked you to get up on a blank whiteboard and draw out a visual representation of the business your customer is in, how they make money, how they differentiate themselves in their market, how their role fits into that "machine," and how they are measured, could you do it? I've asked people to do this exact exercise over the years, both sales reps and sales leaders alike and very rare is the case when someone can jump up and do it flawlessly. Yet, the ability to understand your customers and prospects at this level will completely separate you from every competitor. This is the preparation for the conversation that most aren't either equipped to do or willing to do.

When we engage with clients for the first time, this is the very first exercise we do. We call this research and storyboarding phase "NeuroMessaging®" It's game-changing in the preparation phase of the customer conversation. We'll dive in more on this topic later in this chapter.

Sales Doesn't Start with Why

Author and speaker Simon Sinek is spot on: Before anything else, you need to know your why. As he rightly states, "People don't buy what you do; they buy why you do it."

Having a Why Story is right up his alley. You need to understand what influences you before you influence others. But when it comes to communicating with people, you don't start with why.

You start with "who."

If John, our VP of sales character from our previous story, had been an outside motivational speaker instead of the VP of sales addressing a ballroom full of his employees, he likely would have acted differently. His tone and content would have been tailored for that audience. He also would have acted differently if he'd been presenting to a roomful of customers or, still yet, in the conference room of a big potential customer.

All of us act differently, depending on the situation.

We communicate one way with our coworkers, another way with our long-time customers, and yet another in first-time sales meetings. Perhaps the better example would be Christina. Because she was speaking to her coworkers already inside the organization, she could go out on a limb and talk about her attempted suicide—probably something she wouldn't share with, say, a stranger on a plane. (And certainly not with the TSA agent during security.)

The Neuroscience of Audience Adaptation

This natural tendency to adapt our communication based on context isn't just social convention—it's hardwired into our neural architecture. Neuroimaging studies have revealed a specialized brain network called the "mentalizing system" that activates when we need to predict how others will respond to information.[1]

This system, which includes the temporoparietal junction, posterior superior temporal sulcus, and medial prefrontal cortex, allows us to create mental models of our audience's knowledge, beliefs, and emotional state. The more accurately we can model our audience, the more effectively we can tailor our communication to resonate with them.

What's particularly fascinating is that this audience modeling happens largely unconsciously. When asked about their preparation process, top performers often struggle to articulate exactly how they adapt their approach for different customers. They simply "know" intuitively how to communicate effectively with each audience.

NeuroSelling makes this intuitive process explicit and systematic by starting with a deep understanding of "who"—your specific customer is—before crafting your message.

To create change, your focus has to be on the other person. Christina didn't share her story because she wanted a group therapy session. She did it because she wanted everyone in the room to see how transformative the leadership program was.

Studies have shown that one of the most important traits of consistently successful salespeople is the ability to be empathetic—seeing the customer conversation from the other perspective. In the popular *Harvard Business Review* article "What Makes a Good Salesman," the authors concluded from their research that it boiled down to just two traits: empathy and ambition. They wrote:

> *The salesman with good empathy . . . senses the reactions of the customer and is able to adjust to these reactions. He is not simply bound by a prepared sales track, but he functions in terms of the real interaction between himself and the customer. Sensing what the customer is feeling, he is able to change pace, double back on his track, and make whatever creative modifications might be necessary to home in on the target and close the sale.*[2]

Empathy in the moment is good. But you don't need to be a great reader of body language and other nonverbal clues to be empathetic. You can "prime" yourself to be empathetic by doing your homework before you're in front of your customer. That's the kind of functional empathy you need to be able to cultivate.

The more you practice starting with "who," the more naturally your brain will model your customer's perspective before formulating your approach.

Practical techniques for cultivating this neural empathy include:

1. Researching your customer's industry, role, goals and challenges before meetings
2. Creating detailed customer personas based on data and observation
3. Regularly reviewing and updating your understanding of customer needs
4. Practicing perspective-taking exercises to strengthen neural pathways

This deliberate approach to developing empathy is what distinguishes truly exceptional sales professionals from average performers.

NeuroSelling isn't about gimmicks and in-the-moment tactics. It's about a systematic communication approach with your customers as you continually understand them, their roles, and their problems at a deeper and deeper level.

Understand your customer. Start with who.

Problem Solvers Rule the World

Recently, on a plane ride to a speaking event, the TV in the seat back in front of me began to play that all too familiar "safety" video. "Hi, I'm Joe Blow, CEO of SuperAir"—real airline CEO name and brand omitted to protect the innocent—"We've just finished a multimillion-dollar rebrand of our entire fleet . . ."

I immediately tuned out. I don't care about a new paint job on the tail fin. Does not affect my life at all. I'm two weeks behind on email, presentations, staff development, personal development, and, oh! I haven't gotten my daughter Grace anything for her birthday yet.

This may come as a surprise, but your customers **don't really care about what you're selling.**

- They don't care about how many bells and whistles your gadget has.

- They don't care about how many awards your R&D team has won.

- Let's bring it closer to home. They don't care about your quota.

- They don't care about helping you make commission.

- They don't care that your job may be on the line.

- They don't care about making you look good in front of your boss.

They don't care about your problems at all.

What do they care about?

Their own problems, of course.

Yes, that's right, self-preservation orientation strikes again.

The Neurobiology of Problem-Solving Motivation

This focus on problems isn't a character flaw—it's a fundamental feature of human neurobiology. Research on the brain's motivation and reward systems has revealed that we're hardwired to prioritize solving our own problems above all else.

When we face problems, the brain's limbic system—particularly the amygdala and insula—generates emotional signals that command our attention. These signals trigger the release of stress hormones like cortisol and norepinephrine, creating a physiological state that demands resolution. The longer a problem remains unsolved, the more these neurochemicals accumulate, creating increasingly uncomfortable levels of arousal.

This discomfort creates what neuroscientists call an "approach motivation"—a powerful drive to take action that will reduce the unpleasant arousal. When a potential solution appears, the brain's reward centers activate, releasing dopamine in anticipation of relief. This dopamine release creates a sense of anticipatory pleasure that motivates action.

Most importantly, research has shown that this reward response is significantly stronger when solutions address our own problems compared to others' problems.[3] This explains why your customer is neurologically incapable of caring about your commission or quota—their brain's reward circuitry simply isn't wired to respond to your problems.

What would they welcome more?

Someone there to solve the problem that has them losing sleep.

They're worried that their job performance may not be up to snuff. They're worried that someone else is after their job. Or they may be worried about something at home. Their relationship could be on the rocks. They might be having a tough time with their kid. Their parent might be in the hospital.

The Hair-On-Fire Problem: How to Find What Really Matters

When trying to understand your customer's most pressing concerns, it's helpful to think in terms of what I call their "hair-on-fire problem." This isn't just any challenge they face—it's the specific issue that commands their attention, disrupts their sleep, and dominates their thinking.

Hair-on-fire problems share several key characteristics:

1. They create significant emotional distress – The problem triggers strong emotional responses, activating the limbic system and creating a sense of urgency.
2. They threaten identity or status – The problem potentially jeopardizes how the person sees themselves or how others perceive them.
3. They have meaningful consequences – The problem, if left unsolved, will lead to outcomes the person deeply wants to avoid.
4. They feel personally relevant – The problem directly affects the person's goals, responsibilities, or well-being.
5. They seem somewhat solvable – The problem isn't so overwhelming that it triggers resignation; there's hope for resolution.

Research from the field of neuroscience has shown that problems with these characteristics create unique patterns of brain activation, particularly in regions associated with self-reference, emotional arousal, and motivation.[4] When you identify a problem with these qualities, you've found your entry point for meaningful attention.

When you walk into a customer conversation, you have to stop thinking about your problems. "Can I get this person to buy?" "Can I get a meeting with their boss?" "Is this going to be my big sale this quarter?"

Instead, take a page out of Matt's book and walk in their shoes. What's their hair-on-fire problem, and how can you solve it?

And you can't simply call what you sell a "solution."

It doesn't matter what label you slap on the box: It's not a solution unless it solves their problem.

Your problem is that you need to sell something. That's not their problem. That's yours.

When you sit down with someone, you need to know what their self-preservation orientation is. What's threatening them? What are they most concerned about telling their boss they haven't solved yet in their next one-on-one meeting?

The Problem-Centered Approach to Customer Research

To effectively identify your customer's hair-on-fire problems, you need a systematic approach to customer research. Traditional market research often focuses on demographic data or general trends, but neuroscience suggests that problem-centered research yields far more valuable insights for sales professionals.

Effective problem-centered research includes:

- **Decision maker mapping** – Identifying who makes, influences, approves, and implements purchasing decisions, along with each stakeholder's primary concerns

- **Goal/Objective identification** – Understanding what your customer is really trying to achieve—how they describe it, how they measure success, and what they prioritize—gives you a clear target. That target becomes the anchor point for the problem you'll ultimately help them solve

- **Problem hierarchy analysis** – Discovering which problems command attention based on emotional intensity, perceived urgency, and potential impact

- **Problem validation** – Confirming that the problems you've identified are genuinely important to your customer, not just assumptions based on industry generalizations

- **Solution history exploration** – Learning what approaches the customer has already tried and why these attempts failed to fully resolve the issue

This research should happen before your sales conversation, giving you a preliminary model of your customer's problem landscape. With this foundation, you can then use the conversation itself to refine and validate your understanding.

What's particularly powerful about this approach is that it activates what psychologists call the "attitudinal similarity effect"—the tendency to feel more connected to people who share our concerns.[5] When you demonstrate that you understand your customer's problems before they have to explain them, you create an immediate sense of alignment that accelerates trust building.

Frequently, I like to ask salespeople this question, "**What's the number one job of a salesperson?**"

They answer, "To sell more!" "Make quota!" "Get more customers!"

Wrong. That's what's important to you, the salesperson and your boss. Those answers are all *results* of doing the only thing that creates those answers—**solving your customer's problems**. (Remember Papaw's truism: "Problem-solvers rule the world.")

Subconsciously, we all need to start with "who." We tailor our message and presentation to whoever's sitting in the other chair. But you need to be explicit about it. I want you to start doing it intentionally.

The Impact of Inverted Insomnia

Let me give you an example you're probably familiar with and one I'm intimately familiar with. When I walk into a sales meeting, I know what I need our customers to buy: our NeuroSelling program. My business supplies the salaries for me and my staff. Our families depend on our incomes. I need to provide for the people who depend on me, personally and professionally. I need to cover the cost of our properties and upkeep.

But the person on the other side of the table isn't losing sleep over my issues. I am. They're not worried about how they can help me make payroll. Far from it.

What are they losing sleep over? That's the focus. Unfortunately, we tend to invert the focus of the sleepless night to our own insomnia. Ironically, when your primary focus is helping your customer stop losing sleep by solving their critical issues, the resulting impact is you also tend to sleep easier.

The Attention Economy: Why Problem Focus Matters More Than Ever

In today's business environment, customer attention has become one of the most scarce and valuable resources. Executives are inundated with communication—research shows that the average office worker sends and receives around 121 emails every day.[6] On top of that, they spend nearly 23 hours per week in meetings, more than double the time spent in the 1960s.[7] Meanwhile, workplace focus continues to erode; a study from the University of California, Irvine, found that the average knowledge worker is interrupted or switches tasks every three minutes and five seconds.[8] In such a distracted and overloaded environment, cutting through the noise to connect meaningfully has never been more difficult—or more important.

This overwhelming cognitive load has serious neurological consequences. The prefrontal cortex—responsible for strategic thinking and analysis—has limited resources that deplete quickly under conditions of continuous partial attention. The result is a state that neuroscientists call "attentional residue," where the brain struggles to fully focus on any single task because it's still processing previous inputs.

For sales professionals, this means that your window of opportunity to capture meaningful attention is narrower than ever. Generic pitches and company-centered messaging simply don't meet the threshold required to break through this attentional barrier.

However, research in cognitive psychology has identified one type of stimulus that reliably captures attention even under conditions of cognitive overload: personally relevant problems. When the brain's reticular activating system—the neural network that filters sensory input—detects information relevant to a current problem, it automatically prioritizes that information above competing stimuli.[9]

This explains why focusing on your customer's specific problems is more than just good practice—it's a neurological necessity in today's attention economy.

In our world, our primary customer is a CEO, VP of sales, Chief Revenue Officer, or Chief Sales Officer. What's their worry? What triggers their self-preservation orientation? After working with so many of these folks for so long, I can tell you their top three or four typical objectives:

Increase top-line revenue Decrease sales cycle time Decrease the amount of discounting to get new business Decrease sales rep ramp-up time from hired to producing

If they accomplish these objectives, they'll personally make a significant bonus, be in line for stock options, and likely be considered for a promotion, etc. One level removed, it will allow them to do their part in working toward company growth.

But what's their challenge? What's keeping them from reaching those goals? Again, after working with so many for so long, I can almost read their mind when we're discussing this in person.

Their salespeople have trouble communicating the value of their premium-priced product or service, leading to a lower-than-acceptable closing ratio and a sales cycle that carries on way too long. Additionally, their salespeople typically aren't cross-selling into existing accounts for the same reason.

Due to a lack of effective onboarding, training, and coaching, it typically takes a new sales rep about twelve months to be fully productive by company standards, which is way too long.

They are not worried about putting my three kids through college. They're worried about putting their three kids through college.

After I've shared my Why Story, it would feel disingenuous to go straight into what I want them to buy. If I truly care about them—this new acquaintance that I've shared a part of my life story with and who's shared part of theirs—then what I'm there for is to help them solve their problems.

Why Problems Trump Solutions in the Brain's Hierarchy

There's a fundamental neurological reason why problems command more attention than solutions: the brain evolved primarily as a problem-detection and problem-solving organ. From an evolutionary perspective, our survival depended far more on our ability to detect and respond to threats than on our ability to appreciate opportunities.

This asymmetry remains encoded in our neural architecture. Brain imaging studies have shown that problem-related information activates more widespread neural networks and triggers stronger emotional responses than solution-related information of equal objective importance.

For sales professionals, this creates an important insight: customers are neurologically primed to engage with problem discussions before solution discussions. When you attempt to present solutions before fully exploring problems, you're working against the brain's natural information processing sequence.

However, when customers are guided to discover or articulate problems themselves before solutions are introduced, this resistance diminishes dramatically. Their brain processes the solution not as an external imposition but as a logical response to a problem they've acknowledged and validated.

A Day in the Life of Your Customer

I've worked with plenty of salespeople in the wholesale financial services industry. Their customers are usually financial advisors—the person who sits across the desk or the kitchen table from Mr. and Mrs. Smith, trying to persuade them to invest their retirement portfolio in a certain investment vehicle. The wholesale reps are usually great at going in and bedazzling the financial advisors with their product knowledge and illustration navigation prowess, etc. After all, that is their highest level of training, so naturally, they let it flow like a river.

"Okay," I said to one group of wholesalers. "But what's the financial advisor's problem? What's important to them? What are their goals, personally and professionally?"

I could tell they had never really thought about it quite like that before. I'm not knocking this particular group of professionals, but no one had ever challenged them to walk a mile in a financial advisor's shoes.

We did an exercise around **what a "day in the life" looks like for their customer, the advisor**.

"I'll tell you this: I seriously doubt they bounded out of bed this morning, shouting: 'Aw, man! Today's the day I get to meet the wholesale rep from Financial Rock! I can't wait!' That's not how they start the day," I pointed out.

Financial advisors have tremendous pressure to get new clients, all while worrying about losing the ones they have. If they have three prospect meetings, they're worried about what happens if they don't close them. They're worried about what to say to the two clients who left messages the night before saying that they're worried about what the market's doing. Are they going to jump ship?

While all of this is going on in their mind, they have an appointment with a financial wholesaler who's going to dump a load of information and new products that will take the advisor loads of time and energy to digest and determine if they should use that particular investment vehicle or not.

After going through a typical day, I said, "Now, with that mindset, let's go back to the first question: What's important to them? What do they really care about? Are they really interested in learning about yet another life insurance product, mutual fund, or annuity?"

The Adaptation Principle: Tailoring Your Approach to Customer Needs

This deep customer understanding allows you to implement what we refer to as the "adaptation principle"—the systematic adjustment of your communication approach based on your customer's specific needs, preferences, and cognitive state.

Effective adaptation requires balancing two seemingly contradictory objectives:

1. Maintaining consistency in your core message and value proposition
2. Flexibly adjusting how you deliver that message for maximum resonance (Situational fluency)

Research in communication science has identified specific dimensions along which effective adaptation occurs:

1. Complexity adaptation – Adjusting the technical depth and sophistication of your communication based on the customer's knowledge level
2. Focus adaptation – Emphasizing different aspects of your solution based on the customer's specific priorities
3. Process adaptation – Modifying your sales process to align with the customer's preferred decision-making approach
4. Tempo adaptation – Adjusting the pace and timing of information delivery based on the customer's processing style

What's particularly important is that this adaptation shouldn't feel calculated or manipulative. When adaptation stems from genuine empathy and understanding, customers experience it as responsiveness rather than as an attempt to manipulate.

The insights from that exercise catalyzed an entirely new approach for these wholesalers. Instead of presenting a product, they now begin by talking about retention rates for high-net-worth clients. They talk about how difficult it is to present financial vehicles in a way the layperson can understand. Then, they show the advisor how to lead an effective conversation in a simple yet effective way proven to get more new clients to agree to hand over their hard-earned assets to their care.

Instead of focusing on features and benefits, they've reoriented their entire approach to focusing on solving the customer's problems that can prevent them from accomplishing their goals . . .

. . . just as you're going to learn how to do in Chapter Nine.

From Understanding to Influence: Setting the Stage for NeuroSelling

As we transition to Chapter Nine, it's important to recognize that "starting with who" isn't just preliminary work—it's the foundation upon which all effective influence rests. By developing a comprehensive neural model of your customer before attempting to influence them, you create the conditions for genuine persuasion.

The next chapter will show you how to use this understanding to craft the five critical narratives that drive decision-making. Each of these narratives builds on the customer understanding you've developed by starting with "who."

Remember: Your customer doesn't care about what you're selling. They care about the problems keeping them awake at night. When you truly understand those problems—when you can articulate them more clearly than they can themselves—you've earned the right to offer a solution.

9

END WITH HOW: THE NEUROSELLING© STORY-BASED SITUATIONAL CUSTOMER CONVERSATION

"I like a story well told. That is why I'm sometimes forced to tell them myself."

—MARK TWAIN

ONCE YOU'VE DONE the prep work to really understand your customer, their role, their goals & objectives, the typical problems that prevent them from getting to their desired destination etc., it's time to actually have a conversation with them. Recently, one of my partners at Braintrust, who leads our client coaching program, received a call from one of our sales managers at a financial services client.

"Listen to what just happened to one of our reps," the manager said. "He had gone into an office with three advisors, but only one of them had time to meet him. He used a marker and a whiteboard to walk the advisor through the stories you helped us create with NeuroSelling. When the interactive whiteboard story was complete, the advisor erased the whiteboard, called one of the other partners in and said, 'Do that again!' So, the wholesaler sales rep grabbed the markers, headed to the whiteboard, and did the same story again. Amazingly, the second advisor went and got the third one, then sat down and said, 'Do it again!' I have never seen something quite like this approach!"

What did he do? Simple, really. He walked into the office with a solid understanding of what those advisors were really grappling with ("start with who"). Then he simply followed the NeuroSelling "yellow brick road" of storytelling:

The NeuroSelling® Story Based Situational Customer Conversation Framework

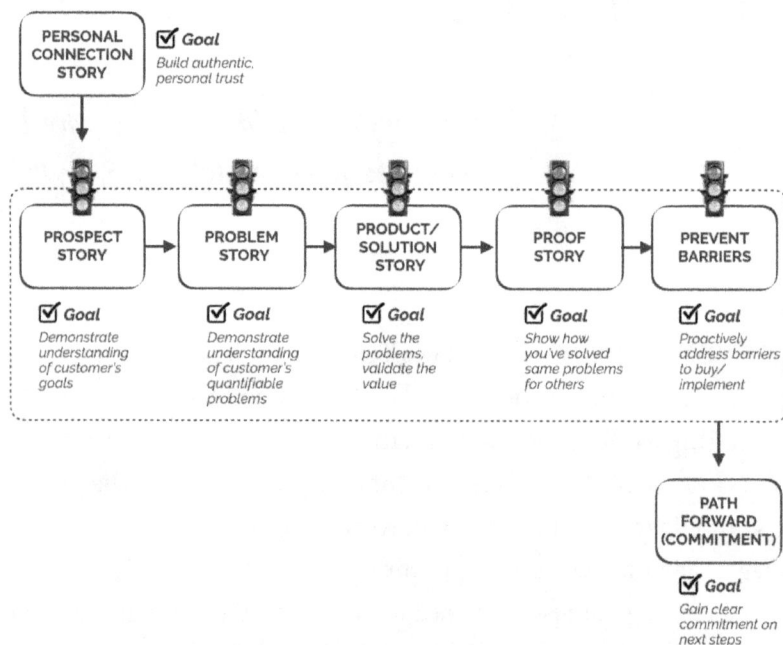

1. Personal connection story
2. Prospect story
3. Problem story
4. Product/solution story
5. Proof story
6. Prevent Barriers
7. Path Forward (Commitment)

The Neuroscience of Sequential Storytelling

This "7 P's" narrative framework isn't just a convenient structure—it's specifically designed to align with how the brain naturally processes information when making decisions. Neuroscientific research has revealed that effective persuasion follows a predictable neural sequence:[1]

- **Trust Formation** → The brain's threat detection systems must be deactivated before meaningful information processing can occur.

- **Personal Relevance Assessment** → The brain determines if information warrants attention based on its relevance to personal goals.

- **Problem Recognition** → The brain's error detection systems identify gaps between current state and desired state.

- **Solution Evaluation** → The brain's analytical systems assess potential paths to resolve the identified gap.

- **Risk Reduction** → The brain's predictive systems seek evidence that the proposed path will succeed.

Each of the steps in the NeuroSelling conversation corresponds to a specific stage in this neural sequence, creating what neuroscientists call "processing fluency"—the ease with which information moves through the brain's evaluation systems. When information follows this natural sequence, it encounters less resistance and requires less cognitive effort to process.

Throughout this book, I've talked about the importance of narratives. This chapter gets to the heart of NeuroSelling: How to take everything we've talked about up to this point and put it together in a sequence that builds authentic trust faster and leads to more urgency to "change" or buy. I'll first give you a high-level overview of each narrative or story, followed by a deeper example of each.

Narrative #1: The Personal Connection "Why Story"

You use this to create a personal connection with your customer and invite them to share why they do what they do. This drives personal trust higher and allows the prospect to lower their "self-preservation" defense shields.

The Oxytocin Effect: How Personal Stories Build Trust

The Personal "Why Story" works because it triggers a powerful neuro-chemical response in your customer's brain. Research by neuroeconomist Paul Zak has demonstrated that character-driven stories with emotional content reliably stimulate the production of oxytocin—often called the "trust hormone."

Oxytocin plays a crucial role in how humans form social bonds and determine who to trust. When oxytocin levels increase, activity in the brain's threat detection centers (particularly the amygdala) decreases, while regions associated with empathy and social connection show enhanced activity.

This neurochemical shift creates what Zak calls a "trust window"—a temporary state in which the brain is more receptive to new information and less likely to engage in skeptical analysis. During this period:

- Credibility assessments become more lenient
- Risk perception decreases
- Empathic connection increases
- Memory encoding strengthens
- Positive attribution bias increases

Most importantly, this oxytocin response doesn't require perfect storytelling. We have seen time and time again that authenticity matters more than polish. A personal story told with genuine emotion triggers stronger oxytocin release than a perfectly crafted but inauthentic narrative.

This explains why starting with your "Why Story" is so effective—it creates a neurochemical environment conducive to trust before you present any business information.

The Personal Connection Story in Detail

Perhaps one of the greatest acceptance speeches I've ever witnessed was Fred "Won't You Be My Neighbor?" Rogers accepting a lifetime achievement award at the twenty-fourth annual Daytime Emmy Awards.

He said, "So many people have helped me to this night. Some of you are here. Some are far away. Some are even in Heaven. All of us have special ones who have loved us into being. Would you just take, along with me, ten seconds to think of the people who have helped you become who you are? Those who have cared about you and wanted what was best for you in life? Ten seconds of silence; I'll watch the time."

After a beautiful stretch of silence—many in the audience with tears in their eyes—our beloved Mr. Rogers continued: "Whomever you've been thinking about, how pleased they must be to know the difference you feel they've made."

Go watch the video. It's worth every second of your time.

The Neurochemistry of Meaningful Connection

This moment from Fred Rogers exemplifies what neuroscientists call "neural resonance"—a state where shared attention to emotionally meaningful content creates synchronized brain activity across individuals.[2] EEG studies of similar emotional moments show remarkably similar neural activity patterns between speakers and listeners, particularly in regions associated with emotional processing and empathy.[3]

What makes this synchronization so powerful is that it creates what psychologists call "emotional contagion"—the unconscious transfer of emotional states between individuals.[4] When genuine emotion is expressed, it activates mirror neuron systems in observers, creating a shared emotional experience.

This neurological connection explains why authentic emotional stories create such deep bonds between people. When you share a meaningful personal story, you're not just exchanging information—you're creating a shared neural and emotional experience that fundamentally changes how your customer relates to you.

Now, I ask you: Who did you think about? Who was someone you looked up to and respected? Who invested some time in your life and made a difference? They're probably your sage.

In my workshops, the Sage has often been a parent or grandparent, but just as often, I've seen others, like an uncle, a foster mother, a band teacher, an older sibling, a minister, or a family friend. It doesn't matter. They only have to matter to you.

On the flip side, a great Why Story is belief-centric, not me-centric.

We don't want to hear about how great you are and how you were the hero of your own tale.

If Christina would have told a story about how she picked herself up by the bootstraps and how strong she was now, her story wouldn't have had nearly the same effect. In her story, her future husband and the leadership program were the real heroes. Matt's story wasn't about him but about his sweet, thoughtful mother and the lessons she instilled. My airplane buddy's story was mostly about his grandpa. My own story isn't about how great Jeff Bloomfield is but how great Papaw Bloomfield was. These characters are the hinges the rest of the story turns on.

I like to call them "The Sage." Just about every great story has a wise old man or woman who helps the hero. Luke Skywalker had Obi-Wan Kenobi and then Yoda. Bilbo and then Frodo had the wizard Gandalf. Queen Ester had Mordecai. Dorothy had Glinda, the Good Witch of the North. The Sage is the person bringing the hero wisdom and inspiring them to do something different.

Character Archetypes and Neural Response

The Sage archetype in storytelling taps into something much deeper than just a good plot—it actually speaks to how our brains are wired. When we hear stories that include a mentor figure, it lights up the parts of our brain tied to moral reasoning and values-based decision-making. That mentor becomes more than just a character—they become a symbol of the kind of person we want to be, helping us connect big-picture principles to the real-life choices we face.

What's powerful about using a mentor figure—rather than casting yourself as the hero—is that it lowers the natural defenses people have when they feel like they're being "sold" or influenced. Our brains are wired to resist persuasion that feels pushy or self-serving. But when you show up as someone who's simply passing along wisdom you've gained rather than acting like you've got all the answers, people lean in. They're more open to the message because it feels like a conversation, not a lecture.

That's the magic of this approach. It shifts the dynamic from "let me teach you" to "let's explore this together." It's no longer about one person having it all figured out—it becomes a shared journey of growth, insight, and wisdom. Let's explore the elements that make up a great personal connection story:

- Element #1: It cannot be about how great you, the narrator, are. The change or call to action has to come from somewhere beyond you. It cannot be me-centric.

- Element #2: It must be belief/values-centric. It isn't really about the narrator; it's about the listener. In my Why Story, I share four beliefs I learned from Papaw: hard work/perseverance, problem-solving, the platinum rule, and the importance of family.

Let me ask you a question: Do you believe all four of those beliefs are important? Likely, you do. I have yet to meet someone who doesn't either share or, at minimum, agree with those beliefs. No one would say that they didn't believe in family or in treating people well, etc. These are universal beliefs.

The Neuroscience of Shared Values

This focus on universal beliefs leverages what we call "value resonance"—the tendency for shared values to create stronger neural synchronization between individuals. When we hear someone express values we share, regions in our prefrontal cortex associated with self-identity show increased activity, creating a sense of alignment and affinity.

Research from Princeton's Social Neuroscience Lab has shown that when speakers and listeners share fundamental values, their neural activity becomes synchronized to a remarkable degree, particularly in regions associated with empathy and trust.[5] This synchronization creates a state where information flows more easily between individuals and is evaluated more positively.

What makes this value resonance particularly powerful in sales contexts is that it operates largely outside of conscious awareness. Customers don't explicitly think, "This person shares my values; therefore, I trust them." Instead, the shared values create an unconscious sense of familiarity and alignment that colors all subsequent interactions.

Remember: To universally connect with people, it's critical to display beliefs that are universally relatable.

Everybody can connect with my airplane friend's belief that you should do what's right, even when nobody's looking. Everybody can believe in Matt's mother's message about walking in someone else's shoes. You don't want to share beliefs that aren't universally accepted, especially the polarizing topics of religion and politics.

Alienating people is the opposite of what you're trying to do.

There's nothing wrong with having core beliefs based around your religious or political affiliations; I do myself. They just aren't very likely to be universal and, in fact, have around a 50 percent chance of actually turning the person against you instead of connecting you. And keep in mind, from a neuroscientific standpoint, this story is designed to allow your prospect to quickly move you from the "foe" or potential foe category to the relatable, possible friend category. It helps you seem less "risky."

Now, think about what you believe. What core values do you truly hold deep down? The kind of beliefs that unite us and form the bedrock of how you've built your life. You may not realize it, but we are united in our beliefs far more than we are different. The problem is that we don't always live out those beliefs, so people are used to being hurt. Getting burned. But when someone reminds us of just how united we are, it restores our own beliefs again and reminds us of the good in others.

Here are just a few common beliefs I've seen in others' Why Stories:

- Choosing forgiveness, especially when it was hard
- Patience being rewarded
- Doing the right thing, no matter what
- Helping others, even if it meant sacrifice
- Doing a job well, even if it meant doing it again
- Doing more than expected
- The Golden Rule (or Platinum Rule, in my case)

- The importance of family
- Being kind, even to a "nobody"
- Problem-solvers rule the world

You have beliefs like this, even if it's been a long time since you thought about them. You expect people to act a certain way, or you believe that there is a right way. When people don't behave that certain way, it probably makes you angry or upset. You believe there is a way the world ought to be.

Figure out what those beliefs are for you.

Now, put Elements #1 and #2 together: How did your sage help you learn those beliefs? It can be directly or even by observing the way they lived their life.

How did you see your beliefs reflected in their actions and life? Try to think about a specific time or a recurring event.

Matt's mother took him and his brother to the homeless kitchen every Saturday—that was what "sacrificing for others" looked like. Him giving away the shoes he and his mother had saved for to a homeless man was "empathy" manifested.

When you think of those specific times, the third element of a great Why Story should come almost naturally—visual anchors.

The Power of Visual Anchors in Memory Formation

Visual anchors are critical to effective storytelling because of how the brain processes and stores memories. Cognitive neuroscience has shown that memories with strong visual components engage multiple brain regions simultaneously, creating what scientists call "multimodal encoding."[6]

When a story includes vivid visual details, it activates not just language processing regions but also the visual cortex, creating stronger and more accessible memory traces. Neuroimaging studies have shown that when people recall stories with strong visual elements, they show activity in many of the same brain regions that would activate if they were actually seeing the events described.[7]

This multimodal encoding explains why visual anchors make stories so much more memorable. Research has shown that information presented with visual elements is remembered up to twenty-two times better than information presented through text or speech alone.

When I'm making my point about hard work and perseverance, I talk about going on the tractor back and forth between the rows of our fields. The visual cue for problem-solving is duct tape. The Platinum Rule looks like my Papaw returning Old Man Crouse's red pickup truck with a full tank of gas. Each belief has a visual that my listener can see in their mind's eye.

When you have your three elements down, the final act is to tie it all together:

The transition: why you're about to tell the listener a story quite different than they're used to. "Before we jump into today's agenda, I'd like to take just two minutes to tell you why I do what I do, then I'd love to hear why you do what you do, okay?"

The scene: transport the listener to the place your story takes place "I grew up on a one-hundred-acre farm in north central Ohio. My papaw bought that farm with his life savings, having moved the family up from Kentucky when my dad was just a boy. It was on this farm that I learned most of the lessons that make me who I am today."

The narrative: tell your story: "I can still remember learning how to drive when I was just five years old . . ."

The bridge: the logical conclusion of how your Why Story is relevant to helping the listener. "So, today, I love helping others learn how to apply these techniques in the way they build and deliver their message. Turns out, it helps just as much at home as it does with prospects and customers!"

The end: an intentional statement to invite them to share their Why Story: "That's why I do what I do. How about you? Why do you do what you do?"

It really is that straightforward.

Simple—just not easy, but once you get it "built," you can customize it for any situation.

The Personal Connection Story – Real-World Example

Here's an example of a Why Story from one of our NeuroSelling graduates at a financial services client, Ben Robertson. I'd like you to read it as if you're the customer and Ben is telling it to you as a way to begin an otherwise high-pressured sales meeting. Then I'd like you to ask yourself, "Do I trust this guy?"

> *Before we get into the agenda today, I'd like to take just a moment to tell you why I do what I do, then I'd love to hear why you do what you do. Growing up in a traditional Southern family, our particular clan was run by a strong patriarch, Otis Kelly Robertson, or, as I knew him, Granddad. Granddad grew up in Hazard, Kentucky. I used to always get a kick out of him telling me the way he gave directions to his house was "head to the first holler, turn at the second creek. The white house with all the chickens." Granddad believed in three things; Faith, Family, and Fishing. All my earliest childhood memories were on the water. Every free weekend, summer vacation, and day off school was spent with Granddad at the lake.*
>
> *He would always have these sayings that I thought were only about fishing, but as I grew up, I learned they were his way of teaching me valuable life lessons. The first was, "You catch your first fish the night before." I can vividly remember him in the garage the night before we were going to the lake, putting new fishing line on poles, tying on different lures, and checking the oil in the boat, all to make sure that when we hit the water in*

the morning, nothing would hinder our ability to immediately start fishing. In other words, the right preparation for something important can be the difference between success and failure, not just in fishing but in life. He also frequently reminded me that "you should always know where all the stumps and rocks are." When you're fishing, hidden structures in the water, such as stumps and rocks, can be your best friend because fish live and hang out near them. But they can also be your worst enemy if you hit one with your boat you didn't know was there while driving down the lake. Knowing your obstacles and learning to use them to your advantage can be the difference in catching your biggest fish of the day or putting a hole in the side of your boat.

He also taught me that "boaters take care of boaters when you're on the water." If you aren't a fisherman, you may not know this, but pleasure boaters (jet skis, tubers, and skiers) aren't always well-liked by the fishing community. When the sun comes up, and it starts to get hot on a summer day, what started out as a quiet, calm cove can turn into a loud and wavy spring-break-style party pretty quickly. Granddad would always just shake his head and curse under his breath as the jet ski flew by way too close to our boat. But one day, while heading back into the boat ramp just before dark, we passed a large ski boat full of younger-looking guys floating in the middle of a main channel, clearly stranded. Their engine had given out, and they were adrift. Granddad didn't hesitate to help. He hooked up a tow rope and towed the boys' boat back into the dock. When you're on the water, all boaters are equal, and we must take care of each other. The financial services industry is like the boating community, and keeping the Golden Rule in mind is paramount. We need each other to succeed.

These life lessons helped shape me into the man I am today and guide my mission with advisors like you every day. Coaching my customers on proper preparation so they hit the water ready to fish, educating them on the obstacles they may run into, and showing them how to turn those same potential obstacles into opportunities while always remembering to take care of others in our industry is why I get out of bed every day. That's why I do what I do.

Now, why do you do what you do?

The Neural Impact of Authentic Storytelling

Stories like Ben's work because they create what neuroscientists call a "neural handshake"—a pattern of brain activation in the listener that mirrors the speaker's brain activity. This neural synchronization is strongest when stories include:

- Concrete sensory details that activate the brain's sensory processing regions
- Emotional content that engages the limbic system
- Character development that activates social cognition networks
- Meaningful values that engage the prefrontal cortex
- Resolution that provides closure and integration

What's particularly powerful about Ben's story is how it leverages what psychologists call "metaphorical thinking"—the brain's ability to understand abstract concepts through concrete experiences. By using fishing as a metaphor for business relationships, Ben creates neural connections between familiar experiences (fishing) and unfamiliar concepts (financial advising principles).

This metaphorical mapping activates existing neural pathways and creates new ones, helping customers process complex ideas more easily. This is why metaphors are such powerful tools in sales communication.

How do you think most of his prospects and customers respond to that story? You guessed it . . . extremely well. They generally launch into their own story about their dad, grandpa, mom, aunt, or other strong influencer in their life who taught them similar beliefs. Please don't think that this type of story is unnecessary. The science would dictate otherwise. It's a fast lane to personal trust.

For those of you who may want to see the elevator-pitch style of this story, here's the condensed version:

> *Before we jump straight into the agenda today, I'd like to take just a minute to share why I do what I do, then I'd love to hear why you do what you do. I was raised in the South and come from a long line of fishermen. My granddad, Otis Robertson, was the patriarch of all patriarchs and the captain of every fishing adventure I went on as a boy. He taught me three things on those adventures that I still value today and believe will have relevance to our conversation in just a bit. First, he taught me that you actually catch the first fish the night before. Success in fishing (and life) can be found in the preparation. The more prepared you are, the better your odds of landing the big one.*
>
> *Next, he taught me that stumps and rocks can be your best friend or your worst enemy. These hidden structures are where the fish tend to hang out, but if you don't know they are there, they can also sink your boat. The lesson? See what others don't see and learn how to take what others see as obstacles and turn them into opportunities.*

Finally, he taught me that you should always be willing to help a fellow boater as you'll never know when you, yourself, might need the help. Who would have thought all those years ago that the lessons I was learning on that lake would allow me to one day help people like you grow their practices, and that's really why I do what I do. How about you? Why do you do what you do?

The first version, the "extended cut," has more detail and can be used when you have more time. That version generally takes between two to three minutes. The second version, the "condensed why," can usually be told in a minute or less. The key for your Why Story is to have a full version and then several iterations that can be used at the right time for the right environment.

Guess what? Your Why Story isn't just for sales. It can also be used as a coach and a leader going for the next promotion! When I worked with Kevin, he was the equivalent of an Executive Vice President or General Manager for the Midwest market of a large enterprise client. Like just about everyone in sales, he was trained from a traditional standpoint: sales was a transaction to be had, not a relationship to be made. But NeuroSelling rang true with him, and he invested his time and energy in learning it.

We worked together for around a year with his team. About six months after our first engagement ended, he invited me to lunch. When we sat down at the restaurant, he said, "This is my treat."

"Thank you!" I said. "Wow, what for?"

He said, "I've just been promoted to CEO. It hasn't been announced yet, but I wanted to share the story with you because my Why Story played a central role in getting me the job.

"The hiring process was pretty rigorous. It came down to me and two other candidates. Really, they were way more qualified and had much more experience than I did. For the final phase, they submitted a list of questions for all of us to answer, followed by a final interview with each of us about our answers.

"Instead of just submitting my answers, I hired a professional videographer and recorded my Why Story. After I told my Why Story toward the end, I went through each of the questions and answered them how they were probably expecting. Then I submitted the video instead of a written response."

In my head, I was thinking, I already see where this is going!

Kevin said, "When we got to the live interview panel, we each had ninety minutes." He paused. "But I only got thirty minutes to answer the original questions."

"What?! Why?" I asked, completely thrown off. I'd expected him to say that it'd gone on for longer—not shorter!

"Well," he said, "because the first sixty minutes was essentially each person on the panel sharing their own personal experiences and talking about how my story resonated with them on such a deep level. Toward the end, they said, 'Oh, yeah, we know you answered the other questions. Just give us a little color around a couple of things.' Then we just ran through a few things, almost like an afterthought. Yeah, everybody told me I had the longest shot out of all three of us, but here I am!"

Later, Kevin found out through the grapevine that the other two interviews had gone very differently, more of a "just the facts, ma'am" interview.

I share this story to point out that NeuroSelling isn't just about selling. Sure, that's a practical application, but it's really a communication model. Whether you're "influencing" someone to hire you, getting a child to eat their vegetables, or showing a customer a new way of thinking, you need to connect with others on a personal level.

The Story Effect in Digital Communication

In today's increasingly digital business environment, the importance of personal storytelling has only increased. Research from the Stanford Virtual

Human Interaction Lab demonstrates that virtual interactions reduce natural oxytocin production by approximately 30 percent compared to in-person interactions.[8] This "digital discount" in trust chemistry creates a significant challenge for remote selling.

Dr. Nick Morgan, in his research on digital communication, explains why: "In the absence of nonverbal cues that typically build connection, the brain searches more actively for other trust signals. Personal narratives provide these signals in abundance, creating what neuroscientists call 'narrative transportation'—a state where the listener becomes immersed in your story despite physical distance."[9]

For sales professionals navigating video calls, phone conversations, and email exchanges, this means your Why Story becomes even more essential. Without the benefit of subtle nonverbal cues that build rapport in person, your personal narrative does the heavy lifting of creating connection.

Building Your Personal Connection Story Library

While your core Why Story provides the foundation for initial connection, effective NeuroSellers develop what I call a "personal story library"—a collection of authentic narratives that illustrate different aspects of their values and experiences. These stories serve as powerful tools throughout the customer relationship.

We teach that you should have at least three or four versions of this story, all with different lengths, to be deployed based on your best situational judgment. Generally speaking, we recommend a thirty-second version, a one-minute version and a longer, more in-depth two-minute version. Begin by building the two-minute version, and then it becomes much easier to trim it down to the other versions.

The key to an effective story library isn't perfection or polish—it's authenticity and relevance. Your stories should reflect genuine experiences that illuminate why you approach your work the way you do, and they should connect to the challenges your customers face.

This single story will change the climate of your customer conversations and create habits that help you avoid "rapport" building and, instead, create a genuine connection.

Narrative #2: The Prospect Story

In the "Prospect Story," you create a story about the prospect that shows you understand their world and have a good handle on their overarching goals and objectives. They are the main character in this story, so gaining agreement and alignment on where they are trying to "go" shows you care about them and understand their business. This will also help establish your professional credibility while, at the same time, you demonstrate empathy for their situation, which further reinforces the personal trust they already have in you that was established with your previous Why Story. In addition, this story serves to allow the prospect to agree, in their own words, to the "target" they are trying to hit.

The Neural Basis of Relevance

The Prospect story works by activating what neuroscientists call the brain's "self-relevance network"—a collection of regions that determine whether information deserves attention and engagement.[10] This network includes the medial prefrontal cortex, posterior cingulate cortex, and temporoparietal junction.

When information is recognized as personally relevant, these regions trigger increased activity in the brain's attention and memory systems. Studies using electroencephalography (EEG) have shown that self-relevant information is processed more deeply and remembered more accurately than non-relevant information, even when the objective importance is identical.[11]

This effect is so powerful that neuroimaging studies have shown that simply hearing or reading one's own name activates this self-reference network. When you expand this to include accurate descriptions of someone's professional goals and objectives, the activation becomes even stronger.

By gaining agreement on the "target" they're trying to hit, you create what psychologists call "implementation intentions"—mental commitments to specific goals that dramatically increase the likelihood of follow-through.[12]

The Prospect Story in Detail

Always keep in mind, from your prospect's perspective, you're walking into their world, their "safety box," and telling them that they need to change. Essentially, you're telling them they've been doing it wrong, but lucky for them, you're there like a masked superhero to save the day!

You're not there to be their knight in shining armor. You're there to help narrate a set of problems they're already likely working through or possibly aren't aware of but should be, and then narrate how the future could look if the two of you partnered together to solve them. Ultimately, your goal is to make them the hero of their own story, not you.

First, we need to make sure that we're talking about the things your customer cares about; this is his or her story, after all. Let's revisit our example whereby we are selling sales enablement, coaching, and training programs. You've already opened with your personal story, so now the "Prospect Story" could go something like this with our executive prospect, Jim:

> Jim, as I was driving in today, I was thinking about all the pressure that must be on your shoulders. I swung into the Shell station to get gas and decided to get something to drink while I was there. I opened the cooler door to the bottles of water section at the back of the store only to be greeted by seven rows of various brands of water, all with different price points. I quickly scanned up and down, grabbed the 'two for one' special, made my way back to my truck, and headed down the road. As I glanced down at my cup holder, I noticed I had a different empty bottle of water left

over from the day before sitting alongside my new bottle from today. The only difference was the label and the color of the lid. Pretty hard to tell them apart. I then realized that on both days, at separate gas stations, I had chosen the least expensive bottle of water each time.

(Pause)

Notice I didn't barge into the room telling the customer what they were doing wrong. Captain Kirk would have instantly raised shields, and Mr. Spock would have begun quickly refuting and negating everything I threw at them. Instead, I told them a story (in this case, using an analogy) they could relate to. Without hard facts to refute, Mr. Spock pipes down, and Captain Kirk leans in, wondering where I'm going with this story. Curiosity and relatability are magical ingredients to every great story. The power is in their hands. They get to come to their own conclusion without me trying to pull them into it.

Creating Multi-Sensory Engagement Through Metaphor

What's happening neurologically during this water bottle story is fascinating. When you use a metaphor or analogy like this, you activate what neuroscientists call "cross-modal processing"—the brain's ability to integrate information across different sensory modalities.[13] This creates a far richer neural engagement than direct factual statements.

Research from the University of London demonstrates that when listeners process a relevant metaphor, they show activation in both the language processing centers (Broca's and Wernicke's areas) and in the sensory and motor regions associated with the metaphorical content.[14] In this case, as you describe selecting water bottles, your listener's visual and motor cortex actually activate as if they were performing the selection themselves.

This multi-sensory engagement creates what psychologists call a "cognitive hook"—a memorable mental image that serves as an anchor for your subsequent message. The water bottle metaphor isn't just a nice story; it's creating an experiential simulation in your customer's brain that primes them to understand the commoditization challenge in their own industry.

> *Jim, I have the privilege of working with several CEOs, presidents, and VPs of sales and marketing. When we discuss their top three or four high-level objectives, it's striking how similar they are. #1: Increase top-line revenue; #2: Improve profitability by reducing the amount of discounting required to secure new business; #3: Shorten the time it takes to get deals completed through the pipeline (sales cycle); and #4: Shorten the time it takes to get new salespeople trained and producing results. Jim, do any or all of those goals resonate with what you are working on today?*

(If we've done our homework, we should never be wrong with these goals, no matter what business we're in.).

> *Jim, tell me what your actual targeted goals are for each—i.e., how much of an increase in revenue are you trying to achieve as a company? How do you currently measure the impact discounting has on your business? What's your average current sales cycle time? What would you like it to be? Finally, do you measure the time it currently takes to get your new reps trained and productive? How much improvement would you like to see here . . . weeks? Months?*

(The above will be a natural, back-and-forth conversation. You are simply getting Jim to tell you his perspective on the "targets," how he measures the targets, etc. This will allow you to properly put those targets at risk with your upcoming Problem Story.)

The sequence is critical: You first establish common objectives that most executives share (creating recognition and validation), then ask for specifics (creating personalization and ownership). This combination creates what neuroscientists call an "attentional spotlight"—focused cognitive resources on particular goals that become more emotionally salient as a result.[15]

I'll take a quick pause here from our Prospect Story to show you what we just accomplished. We used a story to introduce the Prospect Story with the bottle of water narrative. You will soon see where I'm headed with that story.

Suffice it to say, Jim doesn't know either, but he's likely curious.

Next, we mentioned that we work with people just like him (empathy, credibility) and then proceeded to tell him what his peers stated were their typical top three or four goals.

The secret sauce is picking high enough level goals that resonate with your prospect but are aligned in a way that allows you to eventually position a problem that you can solve.

We then asked Jim to help us understand how he currently measures those goals. This gives tremendous insight into how Jim sees his world and the added bonus of staying completely on his agenda.

You have essentially told him a story about himself and allowed him to participate in the details of that story as the main character. When you feel like you're participating in the development of a story, you become even more invested in the outcome being a happy ending! (See choice-supportive bias in Chapter Six). Now that you've set the table nicely with the Prospect Story, it's time to introduce the antagonist to our main character's goals, the "Problem Story."

Narrative #3: The Problem Story

This story is designed to introduce the antagonist or problem(s) preventing them from hitting their "targets" (i.e., goals and objectives) from the previous story. If done correctly, the problem(s) will be associated with the threats and risks to their goals as well as the potential missed opportunities that are either currently or could potentially prevent them from reaching those goals or objectives. Designed within this story is also the need to "quantify" the risk of not hitting the target.

The Neuroscience of Problem Recognition

The Problem story engages what neuroscientists call the brain's "error detection system"—neural circuits that identify discrepancies between desired outcomes and current reality.[16] This system, centered in the anterior cingulate cortex, creates a state of cognitive dissonance that motivates action to resolve the perceived gap.

What's particularly interesting about this system is that it produces stronger neural activation when problems are framed as losses rather than missed gains. Research in neuroeconomics has consistently shown that loss-related information activates the brain's emotion centers (particularly the amygdala and insula) more intensely than equivalent gain-related information.[17]

This neurological asymmetry explains why quantifying the cost of inaction is so much more persuasive than describing potential benefits. When you frame problems in terms of what the customer stands to lose by maintaining the status quo, you create what neuroscientists call a "motivational spotlight"—focused attention on resolving the threat.

The best way to do this is by leveraging third-party insights to underscore the problem. For instance, if you are selling a CRM SaaS solution, you should know that my problem isn't just organizing and tracking the buyer journey and the internal sales process, as that is my goal or objective. My true problem lies in the potential cost of not doing that effectively.

If you told me that according to SBI, nearly 60 percent of deals end in no decision, then began asking me why I think that may be the case, what you should eventually uncover is that due to an inefficient deal-flow tracking system, we are losing over 60 percent of the opportunities in our pipeline. If my average deal size is $20,000 and we are working 100 deals per month but are losing 60 percent of those deals, that's 60 deals lost per month × $20,000 per deal or $1.2 million in potential lost revenue—per month! That's your problem!

Now, how much worse will the problem be if they continue to sit on it? Remember: People don't change until the pain of staying the same becomes greater than the pain of change. You want to ensure they are fully aware of and own the pain of staying the same. This is accomplished by delivering an effective "problem" narrative.

The Loss Aversion Multiplier

This problem quantification process works by leveraging what behavioral economists call "loss aversion"—the tendency to prefer avoiding losses over acquiring equivalent gains. Originally identified by Nobel Prize winners Daniel Kahneman and Amos Tversky, loss aversion is one of the most consistent findings in decision science.

What makes this principle particularly valuable in sales contexts is that the effect isn't linear—it's multiplicative. Research has consistently shown that losses are psychologically about two to two and a half times more powerful than equivalent gains. This means a $100,000 potential loss creates approximately the same motivational force as a $250,000 potential gain.

When people contemplate potential losses, there's increased activity in the amygdala and anterior insula—regions associated with negative emotions and pain processing. This activation creates a physiological stress response that drives avoidance behavior.

By quantifying the cost of inaction, you're not manipulating your customer—you're helping them accurately assess the true stakes of their decision.

The Problem Story in Detail

The easiest action is always inaction. This next step is a crucial link in persuading your customer that they have a problem they really need to do something about.

As we said earlier, your biggest competitor isn't the competition—it's indecision.

Your job is to create urgency to take action by juxtaposing your prospect's current goals against the very antagonist that can or likely may already be preventing them from accomplishing said goals.

At this point in the conversation, it's tempting to launch into the full array of what you offer and how it can help them. But the moment you present anything, Grug Crood is going to rear his ugly head: "This is new. New is bad. Grug no change!"

The moment you introduce anything new, you're telling your customer that they need to leave their safety box. Nobody wants to leave their safety box. Even if it's smelly, dank, and leaks when it rains, we feel safe there.

Before talking about their future (with you), you have to make them understand why their present isn't tenable.

They need to come to their own conclusion that the pain of staying the same outweighs the perceived pain of change.

Grug Crood didn't leave his cave until an earthquake destroyed it. He left because he had to, not because he wanted to. In fact, a smart kid tried to tell him that an earthquake was coming, and Grug wouldn't listen. His family had to have a near-death experience before he finally set out in search of something better. Your job in this next step is to get Grug to see that an earthquake's a-comin'.

Now, remember that you want your customers to have a little spike of cortisol. It focuses their attention on the problem and the insight you're providing.

It's easy: All you have to do is remind them of the fear they already have. Just show them that status quo is not an option.

The Problem Narrative: Creating Controlled Cognitive Dissonance

The most effective problem stories create what psychologists call "controlled cognitive dissonance"—a state where the brain recognizes a troubling inconsistency between current actions and desired outcomes, but in a context where solutions are available.[18] This creates optimal conditions for decision-making.

> *"Jim, I'm one of those nerds who love reading. I read nearly anything I can get my hands on, particularly when it relates to business and important trends that affect my clients. Recently, I was reading the "State of Sales" report that Salesforce puts out each year. It reported that almost 60 percent of sales reps will miss their quota. Further, Sirius Decisions reported that in a survey to sales executives like yourself, the number one reason sales reps miss quota is their inability to articulate value in the customer conversation. This data makes sense as to why CEB found that 86 percent of customers reported not being able to see a difference from one supplier (salesperson) and their competitor. As a result, the Global Chief Sales Officer study reported sales cycles are getting longer in 64 percent of companies!*
>
> *So, back to my original bottle of water story from earlier. Here's the crux of the issue: Companies today struggle to meet revenue goals, discount way too often, and, as a result, elongate their sales cycle because the salespeople can't effectively differentiate themselves and create value in the customer conversation. Period. And as a result of that, your customers view your salespeople the way I viewed those bottles of water in that gas station cooler. Simply the same bottles with a different label. And when you view something as the same, you don't see value and when you don't see value, you either don't choose at all or choose the lowest price. Does any of this seem to resonate with your world today?*

The Authority Principle in Problem Framing

This approach leverages what Robert Cialdini identified as the "authority principle"—our tendency to give more weight to information from credible, knowledgeable sources.[19] By citing specific research from reputable organizations (Salesforce, Sirius Decisions, CEB), you activate regions in the prefrontal cortex associated with credibility assessment, which tend to evaluate third-party data more favorably than vendor claims.

The water bottle metaphor completes the circle by providing what neuroscientists call a "conceptual integration framework"—a concrete mental model that helps the brain organize abstract information into a coherent understanding. This integration creates what psychologists call an "aha moment" —a sudden insight where the problem becomes vividly clear.

Next, to help make the problems concrete, drive just the right amount of cortisol to help Jim focus his attention and create urgency to change, we need to help him "quantify" the real, tangible cost of this problem to him.

> *"Jim, earlier, you mentioned the company has a 20 percent revenue increase goal. What was the average quota for your reps last year?"*
>
> *"$1 million per rep."*
>
> *"Okay, so that means this year, the average quota is around $1.2 million, correct?"*
>
> *"Yes, that's about right, give or take."*
>
> *"And how many reps do you currently have?"*
>
> *"One hundred."*
>
> *"Okay, so you have one hundred reps, all of whom need to increase their sales by $200,000 each, correct?"*
>
> *"Yes."*

"Well, if my farmboy math is correct, it sounds like you have a $200K per rep problem and a $20 million company problem to solve."

The Anchoring Effect in Problem Quantification

This quantification process activates what behavioral economists call the "anchoring effect"—our tendency to rely heavily on the first piece of numerical information encountered when making subsequent judgments.[20] By establishing the $200,000 per rep figure as the "anchor," you create a reference point against which all potential solutions will be evaluated.

You get the idea, right? By putting a cost to the status quo and chunking it down to a per-line problem, the neocortex isn't comparing your price against the cost of doing nothing (which we always think is zero dollars). Instead, you've activated the fear/risk of loss. In this case, the risk of losing somewhere in the neighborhood of $200K per rep if every rep stays at the past year's sales performance.

In reality, you and I know performance always happens on a bell curve, so the reality is likely somewhere in the middle. But here's the beautiful thing, whatever Jim tells you is still a much higher anchor than the likely cost of your solution!

What if he says, "Well . . . most of our reps will be somewhere between 100 and 120 percent to plan. Okay, then let's settle in at 110 percent. That means each rep grew by, on average, $100,000. That still leaves Jim with a $100K/rep problem. Unless you are selling a solution that costs over $100K per person, you have created contrast.

Grug's cave is caving in!

We've covered the first two I's: the issue and the impact.

Now, we need to take it one step further and talk about invasiveness: What's the broader cost of the status quo? Is the problem worse than they think?

Most customers—and, subsequently, most salespeople—focus primarily on the impact the issue/problem has on their direct job or department. That's a myopic view. Every part of an organization is linked. Every action has a ripple effect.

The Invasiveness Element: Expanding Problem Perception

I like to invite the customer to think more broadly:

What's going to happen if things don't change? What happens if production continues to bleed $20 million? What's the impact to other departments if the sales team doesn't hit its quota?

In other words, how invasive is this problem?

By asking them what the broader organizational implications are, it underscores the urgency of the situation and just how big of a problem it really is.

And again, these are their words—not yours.

For example, if they don't hit their growth goals, what gets cut? Where does the leadership find the extra profitability? Do they lay off people? Cut R&D? Reduce bonuses? All of these are invasive "ripple" effects of the problem from an organizational standpoint.

Now, let's continue the "quantification" of the problem to further drive urgency to change.

"When it comes to the discounting problem, what's your current average deal size?"

"$10,000."

"And what's your average discounting rate across the sales org?"

"Typically, we end up seeing around a 10 percent discount to get the business."

"Okay, so again, if my math is correct, If my sales goal is $1.2 million, that's normally 120 deals at $10K per deal. However, if I'm discounting 10 percent, I'll need to sell twelve additional

deals just to produce the same amount of revenue. Essentially, discounting alone is another $120,000 problem. Did I do my math right?"

"Wow, you really do think about this from every angle, Jeff!"

"I know what it's like to be in your shoes, and I know that even small, positive changes can be the difference between making the number and being a hero to the board of directors or missing the number and being the goat."

You see, you don't need to hype up the problem. If you walk them through the scenario to its logical conclusion, they can see for themselves the true gravity of their problem.

Now, let's take another break to review.

At this point:

You've told your Personal Connection Why Story to create personal trust. You then crafted and delivered a compelling Prospect Story and allowed them to participate in the creation of that story. You then pivoted using third-party insights to deliver a compelling Problem Story that puts their current goals and objectives at significant risk. By then quantifying the actual cost of the problem, you now have a significant "anchor" point to work from later when you present your solution and price.

In the end, if salespeople can't drive urgency to change with value clarity in the customer conversation, then you'll end up discounting far too often, elongating the sales cycle, and ultimately miss your revenue goals.

Wow.

That's a lot of ground to cover, but we've allowed Jim to feel in complete control of the agenda since we've only discussed the things he cares about and the problems that could prevent his goals.

The key to remember here is we didn't open the call up with the dreaded "twenty questions." We came armed with knowledge about Jim, we controlled the narrative, and yet Jim still felt in complete control!

Now that we've established the primary problem that needs to be solved for Jim and his organization (sales team effectiveness), we can move on to story number four, the "Product/Solution Story."

Narrative #4: The Product/Solution Story

This is where you solve the problem with your solution. As their partner, the best thing you can do to serve them is to actually help them solve their problem. It just so happens you're doing it with your own product or service. While delivering that narrative, you want to present why you're uniquely qualified to solve their issue. How do you describe your product or service in a simple and understandable way? How does your product or service work? What makes it different? How does it directly and quantifiably solve the problems you've been discussing? Answering these questions will help your prospect see the value of your solution with clarity and simplicity.

The Neural Mechanics of Solution Processing

The Product/Solution story engages what neuroscientists call the brain's "reward prediction" system—neural circuits that evaluate potential paths to desired outcomes.[21] This system, centered in the ventral striatum and orbitofrontal cortex, becomes active when we envision positive future states.

The most effective solution stories create what neuroscientists call "episodic future thinking"—the ability to mentally simulate a better future state in detail.[22] When customers can vividly imagine implementing your solution and experiencing the benefits, their motivation to act increases dramatically.

The Product/Solution Story in Detail

Alright, here's where you get to show off your vast reservoir of credibility. You get to use all the features and benefits you've worked so hard to memorize, the stuff all your presentations and pitches are made of! Finally!

Right?

Of course not.

We're using neuroscience here. We know that the brain can only process so much information before it gets overwhelmed and shuts down. Your customer doesn't care about your product or service. They care about being the hero of their own story. So, paint that picture for them and nothing else.

Don't try to unload all the bells and whistles and everything you could possibly do for them.

Those conversations come later. Right now, you need to be laser-focused on bridging the gap between where they are and where they want to be.

The Cognitive Load Theory in Solution Presentation

This focused approach aligns with what cognitive psychologists call "cognitive load theory"—the understanding that the brain has limited working memory capacity and can only effectively process a small amount of new information at once.[23] Research has shown that when this capacity is exceeded, comprehension and retention drop dramatically, and the likelihood of decision deferral increases.[24]

What's particularly important is that this cognitive overload triggers what we call "decision fatigue"—a state where the brain begins to avoid decisions entirely rather than process excessive information. This explains why feature-heavy presentations so often result in "we need to think about it" responses.

In our case at Braintrust, we have programs that leverage the science of customer decision-making to help salespeople develop trust quicker and create customer conversations that drive urgency to change. That's NeuroSelling in a nutshell. And it's just what the doctor ordered for a client like Jim. Let's dive into the "Product/Solution Story."

"Jim, I understand that changing the way your salespeople communicate can seem like a Herculean task. After all, over the

years you've likely tried any number of various sales training and coaching programs, correct? Do you have any idea why most of them never seem to stick? Usually, it's for two reasons. Number one, these programs typically focus on helping you talk more about you. Guess what? Your customer doesn't care about you. They care about themselves! And two, in order to effectively drive willingness to change in another human being, your customer, in this case, you have to understand the science of human decision-making. You have to communicate the right information at just the right time and in the right order.

"You see, the past couple of decades have brought about neuroscience research that has illuminated how the human brain processes information in order to make a decision. Turns out, the vast majority of sales and marketing professionals communicate with their customer's brain in a way that's not only counterproductive to the way it likes to process information but also their message typically drives skepticism, defensiveness, and doubt—the exact opposite of what you're trying to accomplish! Now that we've cracked the code on the decision-making process, we can use biology, psychology, and physiology to create and deliver messaging that drives trust faster while, at the same time, creates urgency to change. Does this sound like the type of program you'd like to hear more about?"

From here, you can go into as much detail as Jim would like to see relative to how the program works, how it gets implemented, the supporting coaching, how it gets measured etc.

With the problem defined by him and quantified by you, there's just the iceberg question: What barrier exists today that would prevent him from doing something innovative and creative to ensure his salespeople hit or even exceed the goal? His answer becomes the frame for the solution you're about to offer.

For Braintrust, that narrative might go something like this:

"Jim, our sales enablement program—NeuroSelling—takes a scientific approach to the customer conversation. By following our methodology, your reps will connect quicker, build trust more effectively, and create an urgency to change on the part of your prospects.

Today, your reps are likely following a model similar to this: build rapport, ask probing questions to uncover pain, present a solution, handle objections, and close. The problem? This model forces the conversation into the wrong part of your prospect's brain. In fact, it actually speaks to the skeptical brain.

What if you could teach them how the buying brain works and then give them a repeatable model to create clear differentiation from your competitors every time?

The NeuroSelling method uses the knowledge of the buying brain to build a customer conversation around connection first, then credibility by using insight and visual narratives to create clear, compelling differentiation and urgency to buy. We take advantage of the six "limbic levers" of emotion, visualization, experience, contrast, simplicity, and egocentricity to ensure the narrative your reps deliver paints a clear buying vision and a need to solve the problem at hand immediately.

This way, reps see an immediate impact in their ability to create trust and, as a result, see their closing ratios go up and the need to discount go down. As you can imagine, this significantly impacts revenue and shortens sales cycles as well."

The Neuroscience of Solution Presentation

When you present solutions in this format, you activate what neuroscientists call "episodic future thinking"—the brain's ability to mentally simulate future scenarios with vivid detail.[25] This mental simulation creates what psychologists call "pre-experience"—the ability to emotionally experience benefits before they've actually occurred.[26] When customers can vividly imagine implementing your solution and experiencing positive outcomes, their motivation to act increases significantly.

By focusing on how the solution enables the customer to achieve their goals (rather than on the solution's features), you create stronger activation in these motivation centers.

Also, note that the narrative isn't about what Braintrust was going to do for him. I didn't want to engage his skeptical neocortex. I wanted to speak to his welcoming limbic system and root brain. I want him to paint the picture in his mind. I want his brain to picture himself as the hero.

I want him to sell himself.

Now that we've properly peppered the root brain and limbic system, we have to allow Jim's neocortex to participate and "validate" the way he feels in order to justify moving forward.

Narrative #5: The Proof Story

With the "Proof Story," you want to give information for their neocortex that validates what their limbic and root brains already feel. Present a narrative about customers like them who used your product and experienced the results.

Social Proof and the Mirroring Brain

The Proof Story leverages what neuroscientists call "mirror neurons"—specialized brain cells that activate both when we perform an action and when we observe someone else performing that action.[27] These neurons create a direct neural simulation of others' experiences, allowing us to mentally experience what they experienced.

What makes this system particularly powerful in sales contexts is that it's enhanced when we perceive others as similar to ourselves. Research using EEG has shown that mirror neuron activity increases by up to 40 percent when observing actions performed by individuals with whom we identify.[28]

This neural mirroring explains why customer stories about "people like me" are so much more persuasive than generic testimonials or statistical evidence. When customers hear about someone similar to them succeeding with your solution, their brain literally simulates that success experience.

The most effective proof stories activate what psychologists call "transportation"—a state where listeners become so absorbed in a narrative that they temporarily experience decreased critical thinking and increased emotional engagement.[29]

These customer testimonials or "validation" stories ensure the prospect that they aren't alone. It reduces the cortisol and minimizes the potential perception of risk associated with change.

The Proof Story in Detail

I can't tell you how many customer testimonials and case studies I've read that were just a collection of facts and figures: "Customer X used our services. They achieved a reduction in cost of $Y or Z%. Hooray us. It could have been so much more effective structured as a simple narrative.

Again, the neuroscience we're relying on with the Proof Story is that we know the neocortex needs to "activate" at some point in order to justify and validate the feelings we've created in our first four stories. To do that, we are going to initially bypass the neocortex and activate the limbic system and root brain with the story. Since the narrative we'll be using is happening to someone else, there's no threat to the prospect. You're not asking their neocortex to accept or reject anything you're telling them. A story doesn't invite judgment and scrutiny like facts and figures do. You're not asking them to change or do anything, so their self-preservation orientation doesn't trigger.

When I start reading or hearing a properly constructed testimonial story, my brain immediately pictures what's happening, automatically conjuring images and scenes as my internal visualization mechanism starts up. Without even being aware of it, I create an emotional attachment as my brain attempts to empathize with what I'm reading. It's why a great book can make us cry or feel fear: We're in the story with Wilbur as Charlotte spins her web or Frodo as the Nazgûl hunt for him and the ring.

Customer validation stories are made for the NeuroSelling methodology. What if instead of a dry, factual recount of what happened to a client or customer of yours, it was couched in narrative form? That's how we'll deliver Narrative #5: The Proof Story.

> *"Let me take just a moment and tell you about my friend Larry. I'll never forget the first day I met with him. The air conditioning in his building was broken. It was July, so it was ninety-five degrees in his office. We ended up going out to the patio by their cafeteria just to get some air. We were both sweating profusely, and it wasn't because either of us was nervous!*
>
> *After we shared our Why Stories, he confided in me that they were behind again in their sales numbers and, as the VP, he was starting to feel the heat. He had risen to the rank of VP from a sales rep himself and just couldn't figure out how to get others to sell the way he used to. They had tried several tactics from internal boot camps on selling skills to even bringing in an outside consultant last year, but nothing seemed to be sticking. As it turned out, Larry needed to increase each rep's top-line output by an average of $200,000. That may not sound like a lot by itself, but multiply that by the 200 reps they had, and you can see just how daunting a $40 million problem was to Larry.*
>
> *I shared with him that he certainly wasn't alone and that the number one obstacle most companies were having to sales growth was their sales team's inability to tell an effective story that showed differentiation and value. The reason was simple but not easy.*
>
> *Salespeople don't understand how the brain actually does things like build connection, create trust, and ultimately make a buying decision. We walked through NeuroSelling just like you and I have today. He was tired of simply holding manager's meetings*

and trying to get them to be better 'coaches.' That was part of it, but not enough.

He had a decision to make, and seeing how minimal the investment was compared to the potential that this program could have on his sales team and his sales culture, it seemed like the perfect solution.

We engaged his sales team the very next month, starting with the managers, and over the course of the next six months got all 200 reps through our program.

One year later, the sales team had posted a collective $50 million increase in sales, year-over-year. Not only did we help Larry hit his number but we also exceeded it by $10 million!

Needless to say, today Larry is one of our raving fans and has even embedded our program into the rep onboarding training, and he's seeing faster productivity than ever before."

The Social Proof Multiplier Effect

Research by Dr. Jennifer Aaker at Stanford has shown that stories like Larry's are remembered up to twenty-two times more than statistics, partly because they activate both emotional and logical processing systems sequentially.[30]

The specific details in the story (broken air conditioning, the patio meeting, Larry's background as a sales rep) serve as what memory researchers call "elaborative encoding hooks"—concrete sensory details that create multiple retrieval pathways in memory, making the information more accessible later. The unexpected success ($50 million vs. $40 million target) creates a positive prediction error that activates dopamine release, enhancing both attention and memory formation.

After reading that, do those numbers feel more compelling embedded in a narrative? Is it more convincing and compelling than saying, "Yeah, we worked with this company one time. Got their sales reps to sell an extra $50 million."

No context, no connection.

But when it's time to talk about how much your product or service costs, that's where things get rough. Right? Not when you follow the NeuroSelling narratives laid out in this chapter.

Think about it from our prospect example, Jim's perspective. You know you've tried every "sales training du jour" over the years, and you've gotten, at best, modest results. You also believe in this approach. What would you pay to help bridge the problem gap of somewhere between $100,000 to $200,000 per rep? Would you pay $10,000 per rep? $20,000? Either sounds pretty darn good compared to the cost of the problem now, doesn't it? When it comes to presenting pricing, most salespeople flinch as they slide the "napkin" across the table because they've been beaten down by so many customers in the past. They have a form of posttraumatic pricing presentation syndrome. I don't mean to sound harsh, but it's likely your own fault.

Sorry to be so blunt, **but when you haven't created contrast and shown differentiation with value clarity, why wouldn't I ask you for a discount?** Why wouldn't I compare you to the cheapest competitor on the market? You're a bottle of water, after all. Why should I pay more?

The Price-Value Association in Neural Decision-Making

Research in neuroeconomics has revealed a fascinating insight into how we evaluate price: The brain processes price information differently depending on whether it's presented in isolation or in relation to value.[31] When price is presented without clear value context, it activates regions associated with pain and loss (particularly the insula). However, when price is presented after clear value quantification, it activates regions associated with value assessment and reward prediction (particularly the medial prefrontal cortex).

Dr. Brian Knutson at Stanford University demonstrated through fMRI studies that when participants were exposed to product information followed by price, greater activity in reward anticipation regions (nucleus accumbens) and reduced activity in pain-related regions (insula) predicted purchase decisions.[32] This suggests that contextual framing of value can significantly influence the brain's response to price.

This neurological shift explains why the sequence of your NeuroSelling narrative framework is so critical. By establishing problem cost ($100K to $200K per rep) before solution price, you fundamentally change how the brain processes the price information.

The good news is you're a NeuroSelling salesperson now. You've done such a great job of using the "P" narratives that you've built personal trust and driven significant credibility alongside an urgency to solve the problem. It doesn't matter what you sell. The vast majority of our clients are the premium-priced solution in their given industry.

Using NeuroSelling, you have the prospect wanting to choose you without as much concern over the investment as they know the value they are receiving. What a refreshing place to be.

Now that you have the foundation of the stories necessary to drive change using the NeuroSelling communication approach, over the next couple of chapters, we will help you work through how to ask effective questions throughout the customer conversation, as well as how to remove any remaining barriers to change. Your Jedi communication mastery is almost complete.

The Science of Sequential Storytelling: A Summary

The NeuroSelling narrative-based framework isn't just a convenient structure—it's deeply aligned with how the brain naturally processes information when making decisions. Each narrative targets specific neural systems in a sequence that maximizes receptivity and minimizes resistance:

- Personal Story: Activates mirror neuron systems and triggers oxytocin release, creating the neurochemical foundation for trust and relationship-building

- Prospect Story: Engages the brain's self-relevance network and activates goal-representation systems in the prefrontal cortex, creating heightened attention to subsequent information

- Problem Story: Stimulates the brain's error detection system (anterior cingulate cortex) and creates controlled cognitive dissonance that motivates action

- Product/Solution Story: Activates the brain's simulation and prediction systems, creating vivid mental models of potential future states

- Proof Story: Engages vicarious learning mechanisms and activates identification networks, reducing perceived risk through "borrowed experience"

This sequence creates what we call "processing fluency"—the ease with which information moves through the brain's evaluation systems. When information is presented in this order, it requires less cognitive effort to process, encounters less resistance, and is more likely to result in action.

The neuroscience is clear: It's not just what you say but the sequence in which you say it that determines whether customers will trust you, understand the value you offer, and take action to change.

THE ART AND SCIENCE OF INQUIRY: NEUROQUESTIONING

"The art and science of asking questions is the source of all knowledge."

—Thomas Berger, American novelist

WE ADOPTED OUR youngest daughter, Priya, from India when she was two-and-a-half-years old. After she began to get acclimated to her new surroundings, I was reminded of how powerful the curiosity of a toddler can be. However, it didn't take long to realize that her curiosity was different than my older two.

When Grace and Drew were toddlers, they asked questions that most secure, suburban toddlers ask: "Why does the dog have fur and I don't?" "Why does the moon shine at night but not during the day like the sun?"

"Why do you yell at the TV when you're watching football?" You know, those types of questions.

But with Priya, her questions came from a different place. "Will you promise to sleep with me all night and not get up?" "When will mommy be home? She will be home, right?" Her questions were from a point of survival . . . of fear. As she became more and more secure and realized the safety of her new environment, she slowly began to ask questions that showed more curiosity than fear.

As we've covered in multiple places in this book, as humans, our default setting is self-preservation orientation. Not until we feel safe are we open to new ideas. Even when we think about asking effective questions, we have to understand that since we are under stress, we will tend to operate more with a self-serving perspective, even a survival mentality, than one of empathy and concern for the person to whom we are asking the question. When we are worried about our own survival or, in this case, the survival of our sale, we tend to ask questions that are leading and noticeably on our agenda as opposed to our customer's agenda.

Most executives I speak with—or we work with—recognize the need for their sales team to act as consultants and sell "solutions" instead of products, but many CEOs and sales leaders are actually shocked at how poorly their sales teams execute on what they thought was supposed to be a "customer-centric" selling approach.

Recently, I had a conversation with an executive who was sitting next to me on a plane ride to the West Coast. The subject of sales and sales effectiveness came up. It may not surprise you that at one point, he said: "I can always tell when a rep has been through sales training because instead of launching straight into a product pitch, they launch into a list of questions." Admittedly, he knew all too well that neither was the right approach.

Too often, sales teams trying to execute on what they believe is customer-centric, consultative selling never move beyond the "me" first selling approach of "get the salesperson to ask lots of questions, and then match our capabilities to what the client has said."

So, the sales force sits down and makes a list of questions designed to extract information from their prospective clients, in what amounts to, at least from the customer's perspective, an interrogation. I've sat through many sales calls like this, and trust me it's not only ineffective, it's also counterproductive.

To maximize the power of NeuroSelling, we have to move beyond a simplistic view of consultative or even solution selling.

It's not about grilling the buyer but rather engaging in a *thought-provoking, trust-centered, customer-focused, problem-solving dialogue.*

One that focuses on the buyer's priorities, what's in their business's best interests, what prevents them from accomplishing what's in their best interest, and then helping them evaluate your solution against those challenges. Asking questions is part of this engagement process, but there's a right way and a wrong way to do this. Here are some important question-asking "potholes" to avoid in your customer conversations.

The Science Behind Insightful Questions

Insightful questions work because they bypass the surface level of conscious thought and tap into the deeper, more reflective parts of the brain. These are the regions where our true desires, fears, and values reside, often unexamined. By asking questions that prompt reflection on these deeper aspects, we invite a journey inward to a place where the real answers lie.

From a neuroscience perspective, when faced with a thought-provoking question, the brain engages in a fascinating process. Instinctive elaboration refers to an automatic, subconscious process where an individual naturally expands upon new information by linking it to their existing knowledge, experiences, or emotional responses. This cognitive mechanism is innate, driven by the individual's inherent tendencies and past experiences. When encountering new data, the brain instinctively seeks connections with what it already knows or feels, facilitating a deeper understanding and integration of the information into memory.

Essentially, when you ask a question, that very act hijacks the receiver's brain, and they shut down all other neural processing other than the areas responsible for processing and responding to that single thought. As this focused neural activity increases, neuroplasticity comes into play. This term refers to the brain's ability to form and reorganize synaptic connections, especially in response to learning or experience. Insightful questions, therefore, can literally change the structure of your prospect's brain, making them more receptive to new ideas and ways of thinking.

What does this mean for sales professionals? When you ask emotionally engaging questions framed around customer needs (rather than product features), you're literally activating different neural pathways in your customer's brain. You're moving them from passive listening to active engagement.

Let's examine what happens biologically:

- Statement about product features → activates analytical network only

- Standard needs analysis question → triggers analytical engagement

- Emotionally framed question based on customer values → activates emotional network initially, then appropriately engages the analytical network when needed

Stop the Traditional "Needs Analysis"

Recently, we began working with a financial services company that hadn't seen a ton of traction with their current sales approach. After observing a few client engagements, it was easy to see why. The sellers we observed did a decent job of asking lots of questions and getting back lots of answers, but it felt more like they were going through a checklist. Why? Because they were.

The dreaded "needs analysis" and, as a result, their sales calls felt mechanical, transactional, and self-focused.

While they did uncover some good information about clients' needs, allowing them to pivot and pitch the products they were selling, there was little buy-in from the prospects they were talking to and even less urgency to change.

There was no sense of empathy, shared interest, or that the client had confidence that the seller would be able to help them grow their business by solving specific problems that needed to be solved.

I've observed this scenario with both beginner and experienced salespeople and, believe it or not, even senior executives of Fortune 500 clients. What we find is nearly always the same: When you focus primarily on questions for which you need the answer to position your products or services, you rarely get the information you really need.

Start building a questioning strategy focused on the customer's objectives and challenges.

How do you do this?

Well, first you actually need to know your customer's objectives and challenges. "I know, Jeff, that's why we ask them so many questions—to learn those things. Duh."

You see, this is one of the fundamental mistakes being made in sales organizations across the globe today.

You should never walk into a sales meeting until you have a strong understanding of the person or individuals you are meeting with.

What's their role? What does that role typically have as its top five goals and objectives? How are they typically measured relative to those goals? What are the typical challenges or problems they face at accomplishing those goals that you solve for? If you can't answer those questions before you go to the meeting, you have no business going. The art of effective questions comes in learning more about the prospect's feelings around these areas and creating questions that drive ownership of the problems by the prospect and an urgency to do something about it!

For example, if I call on financial advisors, I know that, in general terms, their top goals/objectives are the following:

A. Make more income.

B. Gain more clients.

C. Keep my current clients satisfied so they not only stay with me but also refer me to their friends.

D. Simplify my business so I can have more balance in my life.

If you work in or around the financial services industry, you'll be hard-pressed to find an advisor worth their weight who doesn't have these as top priorities.

Next, what problems or challenges do they face in accomplishing these goals that you solve for? Once you know these, you are armed with the right information to ask more laser-focused, customer-centric questions.

> **Bad question example:** *"Mrs. Advisor, how do you feel about using annuities to supplement your client's retirement strategy?"*

> **Great question example:** *"Mrs. Advisor, I recently read a survey that stated during the last market downturn, 40 percent of clients who left their advisor and went to a new advisor did so because they felt their current advisor didn't do enough to protect their assets. Obviously, when a client leaves you, it not only hurts your income but also reduces your referrals. The right annuity positioned the right way has proven to be one of the most effective hedges against the dreaded market correction. How do you currently use annuities in your client's retirement strategy?"*

Don't miss this.

The difference is that in *the first question*, I ask a self-focused, product-driven question to learn information so I can tell you why you should use my product, in this case, an annuity, more often. All about me.

In the second question, I framed it around two specific problems and tied it to two specific goals you have as an advisor. First, your goal of client retention and second, your goal of referrals. Subtly, I put both those goals at risk using an insight from an article that I read (and be ready to source that article on the spot). Now that I have your mind focused on your goals and then your problems, I land a question that should evoke a more limbic-centered, emotional response.

The difference here is I didn't show up to a meeting with a list of "needs analysis" interrogation questions that only benefit me. **I showed up with a really good understanding of your world and then began a strategy of weaving in questions that are emotional and thought-provoking around areas that you already care about.**

This approach takes a different mindset and a willingness to learn and understand as much about your customer as you do about your product or solution. If you take the time to do this, you will never use a traditional needs analysis again.

The Art of Sequencing Questions

Looking at this through a neuroscience lens, the "order matters" mantra continues to be the expert communicator's calling card. Starting with broad, positive questions around their goals and objectives—you know, the things they actually care about—creates psychological safety. It allows the customer's brain time to reduce cortisol levels and increase oxytocin before you move to more challenging topics.

The optimal question sequence follows this pattern:

1. Connection questions – Build true human connection and establish personal trust

2. Context questions – Understand their broader business landscape

3. Challenge questions – Identify specific problems in relation to their goals

4. Consequence questions – Explore the impact of those problems

5. Clarification questions – Deepen understanding of specific details

This sequence respects how the brain naturally processes information: from general to specific and from low-threat to higher-risk topics.

Stop Asking Self-centered, Leading Questions

Very few things drive change resistance from a prospect faster than a question that either Captain Obvious would ask—i.e., "If your line failed and it cost you $1 million, would that be a problem for your business?" (Yes, I've actually heard that question asked in front of a prospect) or a question that is quite noticeably about you, i.e., "If I can show you how effectively my software works, would you be interested in seeing it in action?" In other words, "If I could show you something interesting, would you be interested?"

The kinds of questions sales professionals are typically taught to ask tend to focus on drawing attention to client problems, "pain points," and other potential sources of disappointment or dissatisfaction, all in an attempt so the client will then view your offerings as a solution. It is useful to explore the buyer's challenges, but when you ask a ridiculous question with an obvious answer such as, "What's the implication of data-center failure?" it usually backfires. It's counterproductive because buyers immediately put up their defenses and will be skeptical of the seller's intentions.

Remember: *Their brain is in risk mitigation mode, and not until they are sure you are trustworthy are they open to how you may help them.*

These types of questions trigger my cortisol, engage my fight-or-flight mechanism, and cause me to retreat inwardly to my closest place of safety. In addition, these types of questions lead me to feel you believe I'm an idiot and have made incredibly dumb decisions to this point, which has me in the predicament (in your mind) that I'm in. Naturally, I will react defensively, even if you are correct.

Start asking empathetic, curiosity-focused questions that your prospect can clarify and quantify.

When you ask questions that demonstrate a genuine curiosity, empathy, and a desire to understand me and my situation or goals better, it activates a different pathway in my brain.

Instead of my analytical pathway that ends in skepticism, distrust, and an unwillingness to consider new ideas, these types of questions travel down my empathic pathway in my brain, which activates interest, builds trust, and creates an openness to learn something new that can help me.

By this point in the book, you know the science well. The key to asking great questions is using the science as your foundation but using the "art" of communication as the vehicle.

Anchor First, Question Second

The brain, in all its powerful and mysterious complexity, is ultimately looking for simplicity. When it comes to asking great questions, think of your questioning strategy like Google Earth. We can all envision being zoomed out so that we see the entire blue marble of the earth. Most of your questions, unfortunately, happen at this level. Questions like "So, tell me what your top goals and priorities are today, etc." This approach is not only uninteresting; it gives the customer the entire library and database of their brain to respond with. We need a more specific target. Let's try to zoom in beyond the blue marble, beyond the continent, and beyond the country and state, and get right into the zip code of the topic we would like to discuss. Setting an anchor provides context, creates relevance, and helps the brain retrieve information more efficiently. Let's take a look at a few before and after examples, with and without "anchoring."

General (Before): "What challenges are you facing in your sales organization?"

Anchored (After): "In conversations with sales leaders across mid-market and enterprise organizations, I keep hearing three recurring themes: 1) Pipeline inconsistency, 2) Reps struggling to articulate true value beyond product features, and 3) Sales managers who are more like super reps than actual coaches. Do any of those resonate with your team, or is there another challenge that's taking more of your focus?"

General (Before): "How do you define success for your team this year?"

Anchored (After): "Many executives I speak with are looking at success in terms of three key buckets: 1) Revenue growth, 2) Retention of top talent, and 3) Increasing customer advocacy and referrals. When you think about your own success metrics, do those three show up in your thinking—or is there a fourth or fifth one that trumps those right now?"

General (Before): "What kind of training or development are you doing with your team?"

Anchored (After): "With everything shifting so fast, most leaders I talk with are investing in three main areas: 1) Helping reps have more consultative, value-driven conversations, 2) Equipping managers to be better coaches, and 3) Building internal alignment around the company's core story. Are you seeing a need in those areas—or is your focus somewhere else right now?"

General (Before): "What makes you different from your competitors?"

Anchored (After): "I've noticed that most companies try to differentiate in one of three ways: 1) Product or tech superiority, 2) Customer experience, or 3) A mission or purpose that resonates on a deeper level. Where do you see your company hanging its hat when it comes to real differentiation?"

General (Before): "Where do you see the biggest opportunity for growth?"

Anchored (After): "Some growth-minded CEOs I work with are focused on one or more of the following: 1) Penetrating new verticals, 2) Expanding wallet share with existing clients, or 3) Launching new offerings that solve emerging pain points. Which of those paths are you exploring—or is there another you'd say is even more pressing?"

When, Where & Which vs. What, How & Why

In the high-stakes world of sales conversations, every word matters—especially the first word of a question. While most sales training emphasizes the importance of asking open-ended questions, the neuroscience of how different question types affect the brain is often overlooked. Emerging research suggests that the specific type of question a salesperson asks—particularly whether it begins with "What," "How," or "Why," or instead "When," "Where," or "Which"—can significantly impact the listener's emotional state, openness, and decision-making ability.

The human brain is wired for survival. At the center of that system is the amygdala, the part of the brain responsible for processing threat and triggering the "fight, flight, or freeze" response. In a sales conversation, certain types of questions can unintentionally activate the amygdala, creating emotional resistance or defensiveness, even if the conversation appears friendly on the surface.

"Why" questions, while often meant to uncover deeper motivations, can easily be interpreted as judgmental or interrogative, especially early in a conversation. For example, asking, "Why did you choose that vendor?" may seem innocuous, but it can subconsciously imply the person made a mistake—activating the amygdala and prompting defensiveness. Studies in communication psychology and motivational interviewing confirm that "why" can feel threatening and should be used sparingly or rephrased more gently.

"What" and "How" questions are generally less threatening, but they often activate the prefrontal cortex, the analytical decision-making region of the brain. While that can be useful in structured problem-solving, it can also lead to overanalysis or suspicion in a sales context if the prospect feels they are being led toward a conclusion that serves the salesperson's agenda. For instance, "What are your top goals?" or "How do you measure success?" may seem helpful but can cause the prospect to retreat into a logical, guarded mindset—especially if the relationship hasn't yet established trust.

On the other hand, questions that begin with "When," "Where," or "Which" tend to orient the brain in time and space, engaging the hippocampus and parietal lobes. These areas are associated with episodic memory and spatial reasoning—cognitive functions that do not trigger the threat response. A question like, "When you think about everything you're responsible for this quarter, which of these outcomes feels most urgent?" is more likely to create a sense of safety and personal reflection rather than pressure or performance evaluation.

These types of questions subtly reduce stress and allow for internal narrative-building. They make the conversation feel exploratory rather than extractive. Neuroscientifically, this approach lowers cognitive load, minimizes emotional defensiveness, and opens up pathways for more meaningful, insight-driven dialogue.

In summary, sales professionals who want to reduce resistance and increase engagement should consider starting with "When," "Where," or "Which" questions to create safety before moving into "What," "How," and especially "Why." By understanding how different question structures impact the brain, salespeople can build more trust, uncover deeper needs, and ultimately influence with more empathy and effectiveness.

The M + S + T Formula for Building Powerful Questions

When building and delivering great questions, it's imperative that you consider three critical areas: Your motivation (M), the question's actual grammatical structure (S) and finally, your tone of voice in delivery (T).

$$M + S + T = Impact$$

Let's think about how this applies to your sales conversations:

Motivation (M): What's your genuine purpose in asking this question? Is it to truly understand the customer's world or just to set up your pitch? The customer can sense the difference.

Structure (S): How is your question constructed? Is it open-ended or closed? Is it simple or multi-layered? Is it framed positively or negatively?

Tone (T): How do you sound when asking? Are you accusatory or collaborative? Curious or judgmental? Your tone can either open doors or close them.

For example, when asking about a customer's challenges with their current solution, consider:

Poor M+S+T: "Don't you find your current system frustrating?" (Manipulative motivation + leading structure + presumptive tone)

Strong M+S+T: "What aspects of your current system work well for you, and where do you see room for improvement?" (Genuine motivation + balanced structure + curious tone)

Stop Triggering Negative Neurochemistry

In her article, "The Neurochemistry of Positive Conversations," Judith Glaser highlights and reinforces much of the science you've been learning in this book. More specifically, she discusses behaviors that contribute to the negative/stress chemical, or "cortisol-producing," and positive/trust chemical "oxytocin-producing" reactions in others.[1]

Among the behaviors that create significant negative impacts are being focused on convincing others and behaving like others don't understand.

These are precisely the behaviors that give salespeople a stereotypical bad name. Behaviors like being too aggressive, not listening, and going on and on about their product or service.

Conversely, the behaviors that create a positive neurochemical reaction include being concerned about others (empathy), stimulating discussions with genuine curiosity, and painting a picture of mutual success (visual storytelling in the limbic system).

Folks who have had tremendous success with NeuroSelling apply these techniques in their discussions with prospects and clients to create a collaborative dynamic with positive outcomes.

As you might have imagined, the degree of trust I have with you will determine how much information I will give you based on the question you ask.

Without trust, I will always be hesitant to give you information that I feel you may use against me, but if I really trust you on a personal level, then I feel safe with you. I feel less risk from answering your questions because I believe you actually care about me; we have a connection, and, in the end, you are the type of person who will do what's right. This is why your Why Story is such a critical tool in your toolbox. It not only helps you connect faster, but it also allows you to ask deeper, more insightful questions in a way that gives you more information in order to help your customer more effectively.

For the remainder of this chapter, I won't bore you with a long list of the different types of questions, from open-ended to closed to probing, etc. etc. etc. Instead, I'd prefer to walk you through the types of questions that work really well within the construct of NeuroSelling and your "P" stories: Personal, Prospect, Problem, Product/Solution, and Proof.

NeuroSelling Narrative-Friendly Questions

In your Personal "Why" Story, asking questions based on how the other person "feels" about their beliefs, their sage, and their purpose can be very powerful.

What I have found is wrapping your story and bridging like this seems to work well, "So that's why I do what I do. How about you? Why do you do what you do?" You might also bridge with a deeper question like, "Do you have someone you can point to that significantly shaped who you are today?" This question is obviously much more personal, but in the right setting can really open up the personal trust floodgates.

Notice I didn't ask them about their resume or their experience. I ask them about their "why." This forces them to think differently about the question and about the story you just told them. As they begin to open up to you about their personal story, ask them questions along the way to demonstrate empathy and understanding.

"Your dad seemed like an incredible guy. If he were still here, what do you think he'd say he's most proud of you for?" or "Wow, that's an incredible story. How do you think the beliefs your mom taught you have helped you be so successful in your career?" These are simple questions, but they drive oxytocin and reinforce trust.

Once you move into the Prospect story, this is where you can begin a strategy of insightful, multilayered questions.

It's important to give your prospect an agreed-upon target first. We can use the one covered earlier in this chapter, "In my experience, most financial advisors I work with are looking to attract new clients, keep their existing clients happy, and grow their revenue/income while simplifying their processes. When you think of your top goals, are those similar to yours, or would you add or subtract from that list?"

Again, simple question, but it's based on an anchor point of goals that I know later I can use my Problem story to put at risk if they don't change. From a multilayered questioning approach, you then drill down further. "When it comes to attracting new clients, what has been your best strategy? What has been the most difficult aspect of gaining new clients?"

Notice that I ask him a "success" question first. I want him to tell me what he thinks he's good at, then I ask him what he struggles with. I can continue this multilayered questioning approach until I feel he's expressed his thoughts and feelings around the goals he's actually trying to accomplish.

Next, in my Problem Story, I will introduce the problem in a unique way. Maybe it's an analogy or metaphor, or maybe it's a traditional story. Either way, it will end in a provocative question. Using third-party insight to position your Problem Story is quite effective.

For example, if I told the prospect a story about a time I went mountain climbing without a guide and did it "free solo" with no anchors or ropes to protect me, I'd certainly have his attention. Then, when I tell him I almost fell a hundred feet to my death, he'll be hanging on every word.

"You know, it really reminds me of the fear investors have of an inevitable stock market correction. Free solo is the fastest way up a mountain . . . kind of like investing in traditional vehicles like stocks and mutual funds can be the fastest way to great returns in a great bull market. The problem is, when the wind picks up and the storm comes, if you're doing "free solo"-type investing only, it's also the fastest way down the mountain. For your clients, if you have all their money tied up in the market and the inevitable market correction hits, what's going to stop their fall? According to a survey by "XYZ" research firm, 75 percent of retirement-aged investors not only said they fear a market correction, but they also fear it will prevent them from living the retirement they had planned. "What is your current strategy to protect your clients from this type of inevitable fall?"

You may feel like, "Boy, Jeff, that sure seems like a long way around asking such a simple question."

Yes . . . and no. By building urgency with the story, you are able to relate the analogy of the free solo "risky" approach without the prospect knowing that's where you were going. It drives curiosity and intrigue.

Then, when you layer in the insight (75 percent of their clients say . . .), you create urgency around the problem. Now, when I ask him what his strategy is to protect his clients, it's much more visual, personal, and urgent. From there you can ask more multilayered questions to help uncover his true strategy and how you may be able to help him.

When you do finally pivot to introduce your Product/Solution story, you should start with a great question that begins to tie everything together. "What if you could offer your prospects something so unique that it not only drove more business but also increased your referrals as well?" Of course, the answer is, "That would be awesome!" This may sound like a "Captain Obvious" question, but it's actually tied directly to their main goals and objectives and is generally rhetorical in nature.

Then, you tell your Product/Solution Story in a way that ties directly back to how it solves the problems they agreed they had earlier in your conversation.

The questions from there should revolve around ensuring the prospect sees and understands the "value" of your solution. "If you began using my recommended strategy (Product/Solution), how do you feel it will help you in both the short and long run?" "Do you believe it will be easy to quantify the value?"

Using AI to Enhance Your Question Strategy

In our digital age, artificial intelligence can be a powerful ally in crafting effective questioning strategies. Modern AI systems can analyze vast amounts of data about customer behaviors, industry trends, and psychological patterns to help you formulate more relevant, impactful questions.

Consider using AI tools to do the following:

- Identify the most common challenges in a specific industry
- Analyze how similar customers have responded to different question approaches
- Generate tailored question frameworks based on customer demographics and psychographics
- Track patterns in customer responses to refine your questioning approach over time

The key is using this technology to become more human in your interactions, not less. AI can help you identify patterns you might miss, allowing you to focus more fully on the human connection during the actual conversation.

When it comes to asking effective "neurocentric" questions, it takes a great deal of awareness, practice, and preparation. You can't just go to "sales training" for a few days and gain mastery of this skill set any more than you can jump on a flight simulator and within a week be able to land an Airbus A320 in the Hudson River like Captain Sully did back in 2009. Putting in the time to craft this type of questioning strategy will go a long way to driving empathy, trust, and an urgency to change on the part of your prospects.

REMOVING THE BARRIERS TO CHANGE

"People who appear to be resisting change may simply be the victim of bad habits. Habit, like gravity, never takes a day off."

—Paul Gibbons, The Science of Successful Organizational Culture

TOM AND HIS wife had already bought the tickets. They'd been planning on a date night out to this show for weeks. At the last minute, though, something came up. You know how that goes.

They offered the tickets to Craig and his wife, Stacey, who they knew not only couldn't afford them but also could really use a night out. No charge, of course. If they couldn't enjoy the performance, they wanted their good friends to have a chance to get away and spend some quality time together.

"Gee, Tom, that sounds great! When is it?" Craig asked.

"This Saturday, seven o'clock," he replied.

"Oh, this Saturday? Man, I don't know. I mean, it's all the way downtown so an hour there, three hours for the show, an hour home, I don't know if we could get someone to watch the kids. Let's see, we'd also need to leave early to have dinner."

On the one hand, Tom wanted to be aggravated. He'd spent a few hundred bucks for these tickets. He could easily go sell them on Facebook in a heartbeat. But he and his wife really wanted to do something nice for their friends. To be honest, he was already slightly annoyed that they themselves couldn't go, but then to have someone they were trying to help start talking about not knowing if they could go because they might need a babysitter or it might be too far away—I mean, did Craig not see that this was the perfect solution to the stress he and his wife were under?

The Moment of Change Resistance

Maybe you've been in a similar situation. Tom's emotional brain was at work. He told himself to calm down and let his neocortex kick in. His friend wasn't necessarily signaling indecision. It did seem he and his wife really wanted to go. He was likely just thinking out loud about all the things that had to fall into place in order to make it happen.

When it comes to our customer conversations, we tend to feel like we have the perfect solution for our customer's issues. Like Tom, we know Craig and Stacey need this night out (our solution) and can't for the life of us understand why he doesn't jump at the chance to implement it, right? Sometimes, the customer "Craigs" of the world just need more information to justify. Other times, there may be some hidden information we aren't aware of, causing him to hesitate. Either way, it can feel frustrating but how we respond can make the difference between deepening the trust we have with the customer versus pushing them further into indecision.

If you've used NeuroSelling as the basis for your sales conversations, you'll be head and shoulders above "the other guys" who continue to fall back on their traditional sales techniques as they stay inside their safety box.

But what happens when you have that moment of change resistance? Traditional sales training often refers to this moment as an objection, but an objection, by definition, is a feeling of disapproval or opposition. If you've sold through old-school means by asking a ton of leading questions then dumping all your "watches" out on the table to convince someone to buy, then yes, you likely get "objections."

In the world of NeuroSelling, however, you establish personal trust up front, gain alignment and agreement on your customer's goals and objectives, then spend the rest of your time discussing the possible issues or problems that might prevent your customer from doing what they already told you they wanted to do!

When you then show them how you can help them prevent or solve those problems with your solution, why would they respond with disapproval or opposition to something they already told you they want help with?

Now, that's not to say they won't still have some anxiety around leaving their safety box. It's just coming from a much different place. In many cases, they are verbalizing the "barriers" to change they have to address in order to implement your solution. In the world of neuroscience, it's their neocortex now trying to ensure they can justify taking action on how they already feel. **They are looking for ways to say "yes" versus challenging you with an adversarial position of why they are saying "no."** In other words, NeuroSelling experts don't "overcome objections"—they help the customer identify and remove any remaining barriers to change.

The Neural Pathways of Resistance

Recent neuroscience research from the Stanford Decision Science Laboratory has revealed fascinating insights into what happens in the brain during moments of resistance to change. Using fMRI technology, researchers observed that the amygdala—our brain's alarm system—activates significantly when people contemplate changing established behaviors.[1]

More importantly, they discovered that different types of resistance activate distinct neural pathways:

- Logical/analytical resistance – Primarily activates the dorsolateral prefrontal cortex

- Emotional/fear-based resistance – Primarily activates the amygdala and limbic system

- Identity-based resistance – Primarily activates the medial prefrontal cortex and posterior cingulate cortex (areas associated with self-concept)

Understanding which type of resistance you're facing allows you to tailor your approach specifically to that neural pathway. For instance, logical resistance responds best to data and rational arguments, while identity-based resistance requires narratives that allow the person to maintain their self-concept while still changing behavior.

Let's take a different approach to Tom's situation. He and his wife knew that their friends had been struggling a bit. His friend's job situation was not the best. They had expressed challenges with their finances, and everything was really beginning to put a strain on their relationship. (In other words, Tom really knew his customer and the customer's situation well.) What if Tom had taken this approach:

"Hey, how are you and Stacey doing?"

"Well, things are pretty rough right now. You know my job stress, and that's putting strain on our finances, and Stacey and I haven't been on a date in over three months."

"Sorry to hear that. We'd like to help. Do you remember telling me that your girls would love for our daughter, Sophie, to come babysit them again?"

"Yes, of course. They love her, and so do we!"

"Well, how about this Saturday night? And I'm going to do you one better: We have tickets for you and Stacey to the theater downtown. Seven o'clock show. Sophie cleared her calendar and can be at your house by three. That will give you both plenty of time to make it downtown, have a nice dinner, and see a great show."

You see, in many cases, when you know your "customer" well and understand their problem, you can position your "solution" in a way that alleviates many of the change barriers in the presentation of the solution itself. How could Tom's friend say "no" at this point?

The key is to think ahead and then be alright with your customer "processing" some of their fears or lingering barriers to saying yes. How you respond is critical. You have to respond in a way that shows your customer you are still on their agenda . . . caring about what they care about. That mindset will help you process those emotions in a more constructive way when you perceive a barrier to change.

Let's take a look at some common ways to do just that.

Override Your Instincts

First, don't react the way Tom did initially.

Don't let your emotional brain start screeching. He was already frustrated, and for a moment, he felt like his friend was either ungrateful or blind to his own problem, which only added fuel to the fire. Tom made the horrible mistake of making the perceived barrier to change about him, not them.

When a customer starts voicing their worries or obstacles, most salespeople's instinct is to retreat to their safety boxes. For most traditionally trained salespeople, that means using facts to tell the customer why they're wrong to feel the way they feel. Seeing the customer's hesitation spikes our cortisol: Oh, no! They're not going to buy! I'm not going to make quota! I've got to make this sale! Yep. Self-preservation at its finest.

Don't let the biology and physiology work against you.

Overcoming barriers to change isn't about responding emotionally as a sales representative, though that will be your instinct. Your instinct will be to pull out your facts and try to club your customer over the head with all the reasons why they're wrong. That's the last thing you want to do.

What you want to do is uncover the motives and the emotions behind the resistance and then reframe the discussion back to their goals and objectives they stated earlier in the conversation.

Let's say Tom's friend still hesitated even after the second approach. Asking a simple question may help reframe his friend back to the problem at hand.

"Listen, you and Stacey have been under tremendous stress. You told me yourself you've been neglecting your relationship for weeks if not months. This sure would go a long way in showing her how much she still means to you. That she's still a priority."

Yes, that's straight to the heart. But it's straight to the heart of the problem! Now, Tom's friend has to reconcile whether or not his relationship is a priority and whether the lack of connection he and Stacey have been feeling is a real problem or not. I know—I'm using a nonselling example to illustrate the point. But my guess is you get the point. Back to the customer . . .

Remember: *Change barriers are emotional on the part of the customer.*

It's their root brain and their limbic system firing off into the self-preservation mechanism. It's trying to reduce risk. It's trying to minimize making a bad decision. So, by understanding this and following through this process, you'll be able to identify the "why" behind the resistance.

Second, what you're hearing as resistance may just be their neocortex speaking up. Don't be afraid when you see your customer start to do this. It could very well mean that they've already made a decision and just need additional information to plug the gaps. Even if it is a genuine barrier, it at least means the customer hasn't dismissed your solution out of hand. They're processing what it would look like to move forward.

Third, it's important to reflect on their barrier relative to where your conversation may have missed the mark. After working as a salesperson, sales manager, and sales and marketing executive for most of my career, I can tell you that when buyers don't make a decision to change, it's usually due to one or more of the following reasons:

1. Lack of trust – This one hurts because it's personal. You likely failed to make the necessary connection. Way to go, lack of differentiated, commoditized bottle of water with a different label salesperson.

2. Lack of urgency – Usually due to you not positioning the problem appropriately or quantifying the cost of status quo.

3. Lack of budget – This can be real but can also be a smokescreen. If the problem you're solving either helps save money or make money in a tangible way, most buyers will find the money to pay for it.

4. Don't see the value – In this case, they don't believe the price of your solution is worth more than what they might save or gain by purchasing it.

5. Don't believe your claims – This ultimately comes down to trust, but, in this case, they simply don't believe your solution will do what you say it will do. Therefore, they don't see it as a solution to a problem.

6. Don't feel the timing is right – Similar to lack of urgency, sometimes a buyer will tell you they love everything about your solution but have to wait until things "calm down," etc. etc. This simply means they don't see the urgency to solve this specific problem today; therefore you likely missed out on an opportunity to quantify the problem somewhere along the conversation.

The Neurochemistry of "Not Now"

"Not now" activates what we call the "uncertainty circuit," involving the anterior cingulate cortex and the insular cortex. These brain regions process conflicting signals and assess risk. Interestingly, this circuit also stimulates a mild dopamine release—a neurochemical reward for avoiding immediate decision-making.[2]

This explains why customers often feel relieved after postponing a decision. Their brain has literally rewarded them with a small "hit" of feel-good chemicals for maintaining the status quo.

To counter this effect, successful NeuroSelling practitioners need to:

- Acknowledge the uncertainty (validating their feeling)

- Reframe immediate action as the lower-risk option (using concrete examples)

- Create a stronger dopamine response associated with moving forward now (by vividly describing positive outcomes)

By understanding this neurochemistry, you can develop specific language patterns that work with—rather than against—these natural brain responses.

Do any of these sound familiar to you? How do your customers express these differently?

Knowing this brings us to our fourth point—understand that you already have the tools to address any one of these barriers to change.

Your Why Story should create connection and trust . . . if you use it.

As you've read, even delivered imperfectly, it's still amazingly effective.

Remember: *Personal trust is the key to the rest of the conversation.*

If they don't trust you, they're never going to tell you that the real reason they don't want to move forward is because they're in the doghouse with their boss right now or that they're embarrassed that the last "solution" totally bombed.

With the other five reasons, you need to go back to your "P" narratives (prospect, problem, product, proof). If they don't feel the urgency or that the timing is right, it probably means you didn't quantify the gap. If you've properly framed the problem as a $4 million-per-production-line problem, I'll bet they could probably find the money in a capital expenditures budget. If they don't feel the need, then you didn't identify the right issue/problem in the first place.

Here's how I might reframe their barrier:

> *Mr. Customer, I must have misunderstood something from our earlier conversation. You mentioned that your primary problem right now is to reduce the overall energy footprint of your business while at the same time saving $1 million per year in direct energy costs. What different options are you looking at to accomplish that goal?*

I mean, I'd be genuinely confused.

Next, I'd like to introduce you to a tool that will help you identify and remove change-resistance barriers.

As mentioned, a change barrier is always rooted in emotion. That emotion is centered around the relationship a customer feels between their perception of risk versus the perception of value. For all my math friends out there, it's a pretty simple set of equations. The depth of the barrier is equal to the level of risk perceived compared to the perception of overall value:

$$Barrier = Risk > Value$$

$$Change = Value > Risk$$

Once you've identified the change barrier, you can begin to ask insightful questions to uncover the root emotions associated with the barrier and then begin to subtly quantify those emotions relative to the customer's perception of risk compared to their belief in the value of your solution.

Let's look at a few examples. You go all the way through your customer conversation with precision, but in the end, your prospect begins to ask a lot of questions around your price. You can tell they think you are too expensive. This is obviously a pretty common barrier.

In this example, let's say you sell a top-of-the-line manufacturing automation software that allows companies to streamline their production which increases throughput, decreases downtime, and does so with the least possible energy consumption on the market. In your conversation, the customer gave you their current problem areas with throughput, downtime, and energy consumption. You were able to quantify those "problem" areas that totaled around $4 million annually. Your solution will cost around $1 million to implement. Let's now look at the customer's resistance barrier through the equation.

What's the emotion here? It's always rooted in fear, but fear of what? What's the risk in their mind?

For starters, your solution will require a significant amount of training for their floor managers and IT department. Next, it will require a complete change in the way the customer moves inventory and raw materials to the lines. In other words, they will have to change their internal processes. These changes alone create a significant amount of anxiety for your prospect.

On a scale of 1–10, I would put them at an 8 in terms of level of risk perception. On the value scale, they see your $1 million price tag and compare that to two things: 1) The cost of doing nothing and 2) The cost of simply upgrading their current technology, which, though not as advanced as your solution, will only cost them $500,000. In their mind, they would likely put the "value" score of your solution at a 4 on a 1–10 scale.

In our equation, the "R" is clearly greater than the "V."

This means your customer's emotional change-resistance barrier score is a 2. Anything over a 1 usually means no deal. Yes, this is a subjective score, but it gives you an internal compass to direct your conversation.

So what can you do?

Ask insightful questions tied back to your earlier conversation.

> *"Mr. Customer, earlier, you expressed to me that, as a company, you had a handful of critical goals. One of those goals was to cut production costs by 20 percent across the board. The other was to reduce your overall energy footprint by 40 percent in the spirit of the 'green' initiative. Let's look at both of those goals again individually.*

> *"On the production cost-cutting front, how will doing nothing or upgrading your current technology impact that goal? Production costs are really a factor of output versus time, correct?"*

> *"Yes."*

> *"With your current system, even if you upgraded it to the maximum available technology, it reduces downtime by 10 percent but doesn't actually increase overall output proportionally. From a very practical standpoint, you can invest $500K on those upgrades and reduce your problem of $4 million down to $3.6 million.*

> *"So, I suppose you could say that reduces your production cost, but does it actually meet the goal? With our solution, you invest $1 million, and you actually eliminate unplanned downtime entirely while at the same time increasing output by 20 percent due to the efficiency of the automation.*

> *"So not only do you go from a $4 million problem to zero, but you also actually increase your production from $20 million per*

line per day to $24 million per line per day. That seems like quite
a difference. Do you agree with my math?"

Leveraging Social Identity Theory in Change Management

Recent advances in social psychology have given us powerful new tools for overcoming resistance. Social Identity Theory, pioneered by Henri Tajfel and further developed by Stanford researchers, provides fascinating insights into why people resist changes that seem objectively beneficial.[3]

The core insight: people don't just make decisions based on personal gain—they make decisions that reinforce their social identity and group membership.

This explains why customers sometimes resist solutions that clearly solve their problems. If adopting your solution somehow threatens their sense of group identity (e.g., "our team has always done it this way"), they'll resist regardless of the rational benefits.

Effective NeuroSellers leverage this understanding by:

1. Identity affirmation – Acknowledging and respecting the customer's current identity
2. Identity bridging – Showing how the new solution aligns with their core values
3. Identity enhancement – Demonstrating how the change strengthens rather than threatens their identity

For example, instead of positioning your solution as a replacement for their current approach, frame it as an evolution that builds upon their existing expertise and reinforces their professional identity.

From there, you can tackle their barrier around fear of internal processes having to change.

"Mr. Customer, you also expressed concern over how your internal
processes will have to change to implement this solution. I can

certainly understand that fear. Typically, the real concern our customers have is around the direct impact to productivity during the implementation as well as the personnel impact to your team. Does that appropriately describe your concern?"

"Yes."

"Well, the good news is we have mastered the implementation process. We have an entire team of implementation specialists who work directly with your team to design and execute the changeover plan one line at a time. Because you don't run every line 24/7, our team works to implement our solution during the off-peak, normal, slow times when you already have lines nonoperational. If that means we are in there from midnight to 6:00 a.m., then so be it.

"The point is, you won't experience production interruption, and your current team will be a part of the solution from day one, so there will be significant buy-in. Does this address your concerns around both the value and the potential fear of "pain of change" in the implementation?"

Here's the key to any barrier to change. You have to identify the barrier. Identify the emotion behind the barrier. Understand the prospect's perception of risk compared to the perception of value and then use the information you should have already uncovered in your NeuroSelling conversation to help alleviate their concern and reduce their cortisol.

It's important to remember the customer has anchor points that they (and hopefully you) have been setting all along your conversation. If you're executing the NeuroSelling model correctly, you've been setting some of those anchors relative to the cost of the problem.

So, by resetting the anchor point back to the cost of the problem—and, most importantly, the cost of doing nothing—you can go back to the opportunity and remind them what their initial goals were to begin with. Then, you can go back to showing them how your solution solves that problem and all the reasons why they want to say yes.

It's taking the barrier and reframing that obstacle as the opportunity that answers or accomplishes the customer's objective.

The brain associates value based on contrast. If I don't believe my problem is worth X and you're charging Y, I'm never going to pay what you're asking for. If the barrier is price, you haven't been able to create enough value between the cost of the problem and the price you're charging. Now that could be on you, it could be on your conversation with the customer, it could be on your pricing model. More than likely, it's on you as a sales representative. You haven't done a thorough enough job of creating the contrast in the customer conversation.

What if, at the end of the conversation, I give my solution and then the price and the customer says, "Whew! That much? I just don't think I'm going to be able to swing that. That seems steep"?

When that happens, I just look at the information we'd been working off of the entire conversation and say, "Well, let me just make sure I understand what you said earlier. We're trying to solve a $4 million problem. Do you feel like that's accurate?" Now, they should because they're the one who said it.

> *"So, what you're telling me is that you don't believe the $1 million investment in my engineering solution will solve that $4 million problem?"*

By asking them that question, you will uncover their true motives for their barrier.

If they don't really believe that to be true, if they don't believe your solution is as good as you think it is, it isn't about the price. It's that they don't believe your solution will solve the problem. Then, you can have an entirely different conversation around capabilities.

Always keep in mind that barriers to change are emotional in nature. They are tied to a set of anchor points your customer has that are likely not the anchors you want them to have. Reset those anchors, reduce the risk, and reestablish the value of your solution relative to the problem. Reset, Reduce, Reestablish.

Say it with me: **"Reset, Reduce, Reestablish."**

Now you are ready to tackle any barrier to change!

CLOSING VS. COMMITTING

"You can have everything in life that you want if you will just help enough other people get what they want."

—ZIG ZIGLAR

ONE OF THE most memorable times in a sales meeting went like this: I'd made a real connection with the VP of Sales. He wanted to roll out NeuroSelling for his whole team. Of course, he needed the CEO's backing. So, he'd gotten all the executives into the boardroom and asked me to "tell our story" again.

I started out with the same "my why" Papaw story I'd shared with the VP and then moved my new audience systematically through the NeuroSelling model. It didn't take long until we got into the real problem: They needed to increase sales by 20 percent to hit their revenue goals.

"How much would a 20 percent increase represent in real dollars?"

The CEO answered, "A hundred million dollars."

"And how many salespeople do you have?"

"A hundred," the sales VP answered.

"So, you really have a million-dollar-per-rep problem, don't you?" I asked.

We continued to talk about NeuroSelling, the science behind effective communication, and what Braintrust had done for others.

After about forty-five minutes of great whiteboarding and questions going back and forth, the CEO stood up and said, "Jeff, this sounds exactly like what we need. Let's get it on the books!" He shook my hand and then walked out the door.

"Hey!" I called after him. "Don't you care how much it costs?!"

He poked his head back in the room and laughed. "Sure! How much does it cost?"

"Well, you have a $1 million-per-rep problem, right? How much is it worth to you to solve it?"

He laughed again and walked back out.

Over his shoulder, he yelled, "Teach my team how to negotiate like you! Send me the bill!"

I mean, right there, in front of all his VPs, he'd basically just given me a blank check. Of course, we took them through the same program at our standard rate, but think about it: What a world of difference to go from defending your value and competing on price to the customer dying laughing and not even worried about getting the best price?

Once you know how big the problem is—and the cost of inaction—then the solution that you eventually propose should look like a bargain by comparison.

If they have a $1 million-per-rep problem, how much would they be willing to spend to solve that problem?

The Neuroscience of Commitment

Before diving into the practical aspects of gaining commitment, it's important to understand what's happening in the brain during this critical phase of the buying journey. Recent neuroscience research has provided fascinating insights into the distinction between a forced "close" versus an authentic commitment.

We've certainly established throughout this book that when people feel pressured into a decision (as in traditional closing techniques), the brain's threat detection system activates—specifically the amygdala and anterior insula. This creates a stress response that literally makes it harder for customers to say "yes."

In contrast, when customers feel they're making their own choice to commit, the brain's reward centers activate—particularly the ventral striatum and orbitofrontal cortex. This creates a positive emotional state associated with the decision, essentially "rewarding" the brain for making the choice.

This fundamental difference explains why the traditional "ABC" (Always Be Closing) approach is neurologically counterproductive. When we pressure customers, we're literally triggering their brain's defense mechanisms. Instead, by empowering them to make their own choice, we activate reward circuits that make commitment feel good.

NeuroSelling Salespeople Don't Close

You've probably been taught one of two sales methods: either the hard close (yikes!) or to wait for the customer to ask for a contract (yawn). Neither is an option in NeuroSelling.

Gong, a big data analytics sales research firm, has now used their technology to analyze over a million sales conversations. They set out to determine the trends and favorable characteristics that lead to new business. Guess what they found? Closing techniques are not only ineffective, but they also actually cause you to sell less. In fact, they found that the best way to positively impact the end of your customer conversations is to change the beginning!

Chris Orlob, Director of Sales at Gong.io, uses a great analogy to explain their research findings. Imagine an asteroid that has already entered the Earth's atmosphere. Changing its course at that point (toward the end of the journey) will have little effect on the overall catastrophic impact. However, when the asteroid is all the way back at the beginning of its journey, even a slight modification to its path will bring about substantial changes to the overall trajectory of the journey and, hence, the outcome.

So it is with your customer conversations as well. Trying to "correct" or change closing techniques is a waste of your time. Improving your skills earlier in the customer conversation, as we've been discussing this entire book, will yield far greater results for the end of the journey.

Great salespeople don't close. They lead the customer to a natural, logical "choice" to solve their problem with your solution!

Just like with that CEO laughing his way out of the room, I didn't have to close him. I didn't need to close him. If I have built great personal and professional trust and forged a great partnership with someone, working together just seems like the right thing to do. The smart thing to do.

The Science of Choice Architecture

Richard Thaler and Cass Sunstein's pioneering work on "choice architecture" (the way choices are presented to people) provides valuable insights for gaining commitment.[1] Their research shows that people make decisions based not just on what they choose but on how the choices are framed and presented.

If you've earned their personal and professional trust and let them come to the decision themselves, then you have earned the right to ask for their commitment but that "ask" can't be your typical closing questions. You must transfer the power, the choice, back to the prospect at that point. It has to be their choice.

This could begin with something like:

> *"Mr. Customer, based on the conversation we've had today, it's clear that solving these issues together will be mutually beneficial. We will have a great deal of time, money, and resources invested in delivering this solution, as will you and your company, but I believe it will be worth the effort on both of our parts."*

Obviously, this is the windup to the pitch. You're setting them up for the sale.

But then . . .

The Million-Dollar Question

This one question is so important that I've devoted an entire subheading to it. I call it the million-dollar question because it's a million-dollar idea that's made our team and our clients millions of dollars over the past decade. Drumroll, please . . . here's the question:

"When you think about everything we discussed today—from the goals you are trying to accomplish to the problems that we are uniquely qualified to help you solve—

What would you like to do?"

Don't miss the power in the simplicity of that question.

When our children were younger and we were trying to instill independence, we would lay out two options. "We can wait until Sunday morning to lay our clothes out for church. Or, we can lay our clothes out tonight, leave earlier, and have enough time to go get doughnuts. What would you like to do?"

Of course, they wanted to get doughnuts. They willingly and cheerfully got their clothes out. We didn't have to tell them to. We didn't have to sell them on the idea. They saw the options and chose to do it themselves.

And if you think that we adults aren't just big kids, think back to the last time someone ordered you to do something. "Well, isn't that what you were going to do anyway?" "Yeah, but I didn't need him to tell me that!" We don't like having decisions forced upon us.

You can't order your customer to buy from you, but the psychology is the same. If you tell them, "Okay, here's what we need to do to move forward," then you've taken that choice away. You're making the decision for them. Nobody likes that, no matter how old they are.

On the other hand, if you never ask for their business, then you never give them a chance to say yes. I once heard someone say, "If you give a customer all the time in the world, they'll take it." If they don't know that it's time to take action—if you don't ask for some sort of commitment—then how are they supposed to know?

Simply ask: "What would you like to do?"

The Psychology Behind "What Would You Like to Do?"

The power of the "what would you like to do?" approach lies in its activation of what psychologists call "choice-supportive bias" and "post-decision satisfaction." When people make a choice themselves (versus being told what to do), several psychological mechanisms activate:

1. Ownership Effect – The brain's orbitofrontal cortex activates when we feel ownership of a decision, creating a stronger commitment to follow through.
2. Cognitive Dissonance Reduction – The brain naturally works to justify choices we've made freely, reinforcing our commitment
3. Autonomy Motivation – Self-determination theory shows that autonomy is a core psychological need; when satisfied, it creates intrinsic motivation.
4. Status Preservation – Making our own choices maintains our sense of status and control, activating reward centers in the ventromedial prefrontal cortex.

These mechanisms explain why this simple question is so powerful. By asking, "What would you like to do?" rather than "Should we move forward?" or "Are you ready to sign?" you're deliberately activating neural pathways that reinforce commitment and minimize regret.

Gaining Commitment

NeuroSelling devotees don't close. We gain commitment.

Like everything else you've read in this book, I hope you understand that the meaning is deeper than semantics.

At the end of your Product/Solution Story, when you're presenting your solution and tying it back to the problem, we don't believe in the cheesy, "So, what's it going to take to get you to say yes today, Mr. Customer?" nonsensical question because we know that doesn't work.

"If you've done the homework, the test is easy." If you've established connection and credibility and followed the rest of NeuroSelling, then this step should almost take care of itself.

I recommend saying something like, "Mr. Customer, based on the conversation we've had today, the problem you're experiencing, and how much it's actually costing you, it seems like our solution is the perfect fit. So, what would you like to do?"

Now, that may seem like a completely simple and innocuous question, but it's powerful in its neuroscience capabilities. Remember those cognitive biases? When I say, "What would you like to do?" I am giving you the power. You have the choice to take ownership. I'm allowing you to confirm, using your own confirmation and choice-supportive bias in my favor. I'm also allowing you to take the opportunity to jump on the bandwagon and join the other customers who've felt the same way that we want you to feel by saying yes.

I'm putting all the power over to you, but I'm not allowing you to not make a choice. So, by saying, "What would you like to do?" you have to respond with some form of action. How you respond to that form of action will tell me exactly where I stand in relation to getting this deal done.

If you say you'd like to take time to think about it, then I know that I haven't closed the gap. I know that I haven't presented something somewhere along the conversation in a way that is going to drive urgency to change. Great sales representatives don't like to leave a sales conversation with a maybe. I'd much rather you tell me no and the reasons why or tell me yes. But "maybe" is never an option for me.

If I think I've done everything correctly in the first five steps of our model, I've earned the right to get an honest yes or no.

This is what we call the power of the "partnership agreement."

"Mr. Customer, in order for us to implement this solution, you know, we have a lot of time, people, energy, and resources invested in bringing this solution to you, just like you're going to have a lot of time, resources, and energy invested in us solving this problem with our solution. So, collectively, we've got a lot at stake joining together to solve this problem."

Language like this goes a long way to helping create a partnership model in the mind of the customer. Sometimes, they forget that you have a lot at stake in solving this problem. Sometimes, they forget that you have a lot of energy and time and money and resources invested in helping bring this solution to them. Just by simply stating that with empathy and humility, you remind the subconscious part of their brain that they trust you, feel a connection to you, and this is a win/win solution.

Strengthening Post-Decision Satisfaction

An often-overlooked aspect of gaining commitment is what happens immediately after the customer says "yes." Research in behavioral economics shows that people often experience "post-decision anxiety"—a form of buyer's remorse that can emerge even before the purchase is complete.

The most effective NeuroSellers don't just gain commitment—they reinforce it immediately using three scientifically proven techniques:

1. Immediate value delivery – Providing an unexpected "bonus" value immediately after commitment (often information-based) creates what psychologists call a "peak-end rule" memory effect

2. Future-casting – Walking through specific positive outcomes in vivid detail activates reward anticipation in the nucleus accumbens (the brain's pleasure center)

3. Social validation reinforcement – Sharing specific examples of similar customers who made the same choice activates social conformity centers in the prefrontal cortex

By implementing these techniques in the moments immediately following commitment, you're literally "hardwiring" satisfaction into the customer's brain by creating positive emotional associations with their decision.

A Commitment-Centered Approach to Complex Sales

The principles we've discussed take on special significance in complex, multi-stakeholder B2B sales environments. Research from the Corporate Executive Board (now part of Gartner) and later confirmed by Harvard Business Review shows that the average B2B purchase decision involves around 6.8 stakeholders, each bringing unique priorities, concerns, and perspectives. More recent insights suggest that this number continues to rise, with some studies indicating as many as 11.4 individuals now influencing complex B2B buying decisions. This growing number of decision-makers underscores the need for sales professionals to navigate a wider array of stakeholder dynamics than ever before.[2]

In these complex environments, securing commitment requires a strategic approach that addresses the neural and psychological dynamics of group decision-making:

1. Build consensus incrementally – Research shows that seeking a series of small commitments activates consistency bias in each stakeholder's brain, making final commitment more likely.

2. Create collective ownership – Using collaborative language ("our plan" vs. "my recommendation") activates social belonging circuits in the brain's limbic system.

3. Manage status dynamics – Acknowledging each stakeholder's expertise activates status-related reward centers in the nucleus accumbens.

4. Address risk asymmetry – Different stakeholders have different risk perceptions; personalizing risk mitigation for each activates individual relief responses.

When properly applied, these principles create what psychologists call "psychological safety"—a group dynamic where stakeholders feel comfortable committing to meaningful change together.

To summarize this chapter, we have reframed what many still call "closing." Traditional sales methods often rely on pressure tactics, manipulation, or scripted closing techniques. But neuroscience shows us that true commitment is never forced—it's formed.

When we understand how the brain processes decisions, especially in high-stakes or complex environments, we realize that the most effective path forward is one that empowers the customer. A simple, sincere question—"What would you like to do?"—does more than move the process forward. It activates powerful psychological drivers like autonomy, ownership, and status preservation. The customer doesn't feel closed—they feel in control.

This shift is more than semantic. It's strategic. When we invite the customer into a shared decision rather than pushing them toward one, we build lasting trust. And when we follow that up with post-decision reinforcement strategies, especially in multi-stakeholder B2B environments, we don't just win a deal—we establish a long-term partnership rooted in mutual value.

In NeuroSelling, the close isn't the conquest—it's the culmination of a conversation built on trust, science, and shared purpose.

EMOTIONAL INTELLIGENCE: THE SALES SUCCESS AMPLIFIER

"In a very real sense, we have two minds, one that thinks and one that feels. Emotional intelligence is the ability to sense, understand, and effectively apply the power of emotions."
—Daniel Goleman

IT'S EARLY MORNING, and I'm back in my driveway, sliding materials into the trunk of my car—just like I used to. Years ago, we wore suits, walked into offices, and had maybe thirty seconds to make an impression. I didn't have the language for it then, but I felt it: the tension, anticipation, and emotional charge from the front desk to the decision-maker's office. That was Emotional intelligence at work—even if I didn't know it yet.

Today, we understand those moments better. Emotional intelligence (EI/EQ) isn't just about managing feelings—it's about managing impact. In sales, your ability to connect, influence, adapt, and build trust hinges on four core domains: self-awareness, self-management, social aware-ness, and relationship management. And inside those domains? Twelve competencies that consistently show up in top performers. These aren't buzzwords—they're the difference-makers.

Let me bring this a bit closer to home.

My daughter is a performer—young, talented, and full of fire. She steps into audition rooms where the stakes feel sky-high. She preps, she performs, and she walks out. Then she waits. Sometimes, she gets the part. Most times, she doesn't. And yet, she shows up again the next day. That's EI: the ability to ride the emotional waves and stay centered through rejec-tion and uncertainty.

Now, my son. Fresh out of college, a Gen Z professional landed his first job a few years ago at a Fortune 5 healthcare company. He's smart and hungry, trying to prove himself in a high-pressure, performance-driven world. He's prospecting, prepping, pitching, and answering objections— often with very little control over the outcome. The stakes are real. The pressure? Constant. As he works with brokers, processes hundreds of emails, works with internal stakeholders, and works with clients, he is learning the impact of keeping emotions under control, especially as the stress grows.

They're in completely different worlds, but they share the same cycle: prepare, perform, evaluate, win/lose, do it again. The emotions are high, the feedback is uncertain, and the ability to bounce back is non-negotiable. Sound familiar?

As we think about sales, there is no debate that it is an emotional sport.

And the science backs it up. Dr. Richard Boyatzis, one of our Braintrust mentors, alongside Daniel Goleman, helped develop the Emotional and Social Competency Inventory (ESCI). Korn Ferry validated this tool with over 80,000 participants across 2,200 organizations. The result? EI directly correlates with performance.[1]

Let me show you what this looks like in the field.

Take Sarah, a seasoned sales rep working a deal for five months. Pressure's mounting, and the buyer goes quiet. She could panic, but she doesn't. She journals. Reflects. That's emotional self-awareness. She breathes, walks, and centers herself. That's emotional self-control. On the next call, she listens more deeply. She senses a shift in priorities and asks a question that unlocks the truth. That's social awareness. She pivots and aligns the offer with their new needs. That's adaptability and influence. A week later? Deal closed.

But more than that—trust earned.

Then there's Marcus. Walking into a meeting with a buyer known for chewing up and spitting out reps. Instead of bracing for a fight, he visualizes success. Slows his breathing. Walks in calmly, present, and focused. He doesn't win the deal that day. But he wins something more valuable—a second meeting. That's EI in motion.[2]

Through the eyes of these multiple scenarios, let's break down the twelve ESCI competencies through a sales lens:

1. Emotional Self-Awareness – Know your triggers and how your energy enters the room.
2. Emotional Self-Control – Don't let one ghosted email derail your mindset.
3. Adaptability – Flex your approach. No two buyers are the same.
4. Achievement Orientation – Chase excellence, not the bare minimum.
5. Positive Outlook – Stay hopeful, even when the pipeline's thin.
6. Empathy – Truly feel what they feel—and let them know it.
7. Organizational Awareness – Understand the decision dynamics.
8. Influence – Move people, not manipulate them.
9. Inspirational Leadership – Make your message contagious.
10. Coach and Mentor – Lift up others. Great teams rise together.
11. Conflict Management – Navigate tension with professionalism.
12. Teamwork – Sales isn't solo. Rally the village.

Still think these are "soft" skills? The World Economic Forum doesn't. Emotional intelligence has been one of their top ten skills for future jobs since 2020.[3] The Center for Creative Leadership found that 75 percent of career derailments stem from emotional and social skill gaps—not technical ones.[4]

Let's stop calling them soft skills. These are the real skills.

Now, let's map EI into the NeuroSelling framework. At Braintrust, we know the brain makes decisions emotionally before it engages logic.

Personal Connection Story – Requires Positive Outlook and Self-Awareness.

Prospect Story – Built on Empathy and Organizational Awareness.

Problem Story – Demands Adaptability and Self-Control.

Product/Solution Story – Powered by Influence and Conflict Management.

Partnership Path Forward – Depends on Teamwork, Achievement Orientation, and Empathy.

Lead with story. Lead with emotion. That's where decisions happen.

In 1970, Walter Mischel conducted a study on delayed gratification, giving kids a choice: one marshmallow now or two if they waited. The kids who waited had better outcomes—because they had better emotional regulation.[5] Later studies added nuance, but the core insight holds: Delaying gratification is powerful.

In sales, deals fall through, calls go unanswered, and pipelines dry up. But EI keeps you from unraveling. It helps you rebound, reset, and start fresh.

EI even shows up in your body. Emotions trigger neurochemicals like cortisol and oxytocin, impacting your nervous system and your perception.

When you learn to regulate those chemicals, you perform better. But more importantly, you relate and lead better.

So, how do you build EI? Here are some tips that you can immediately activate.

Self-Awareness

- Know your values.
- Journal in moments of stress.
- Identify emotional triggers.
- Notice how your mood affects others.

Self-Management

- Take a breath before responding.
- Visualize your ideal outcome.
- Schedule recovery time.
- Smile. Humor heals.

Social Awareness

- Listen without formulating your reply.
- Use names.
- Watch body language for clues.

Relationship Management

- Have hard conversations with grace.
- Build trust consistently.
- Guide others through emotional moments.

You've worked a deal for six months. You've met the client five times, aligned the team, prepped the RFP, and on Friday afternoon, you get the text: "We're going with someone else." That sting is real. But if you've developed your EI, you let yourself feel it. Then, you release it. Monday brings a new opportunity; this isn't just about you but the mission.

I've seen this in reps, my kids, and the best leaders I've worked with. They're not emotionally robotic. They're emotionally resilient.

So, let's get real.

Your IQ might get you in the door. But your EI keeps you in the room. These aren't optional skills. They're essential. And they're learnable.

Let's break a few of these down even further.

Take Empathy. I worked with a rep named Lila, who sold into hospital systems. She had one tough buyer—logical, curt, numbers-driven. Lila did her homework. She didn't just prep the business case—she learned that the buyer was leading a department undergoing restructuring. On the call, she didn't launch into features. She opened with a simple question: "How are your people holding up?" That moment changed everything. The buyer exhaled. They talked for fifteen minutes before ever touching the deck. Empathy opened the door logic couldn't.

Organizational awareness often gets overlooked. One of our clients at Braintrust was selling into a large manufacturing company. The salesperson kept pitching to the most enthusiastic contact, only to lose the deal every time. We coached them to map the org chart and observe internal dynamics. Who had influence? Who had the final say? The next time, they included a quiet VP early on. That VP ended up being the key decision-maker. Awareness isn't optional—it's strategic. This is about great personal and account management discipline.

Coach and mentor isn't just about your team—it's about your customers, too. A high-performing rep named Omar once told me, "If my buyer can't explain the solution to their boss, I've failed." He didn't just sell; he taught. He armed his champion with clarity and confidence. That mindset created not just sales but advocates. We constantly remind ourselves in prospective future client meetings to maintain emotional control, listen more than we speak, ask real and curious questions, and follow the NeuroSelling® approach. Now, let's paint a picture.

A Day in the Life of a High-EI Rep

Meet Jen. It's 7:30 a.m., and she's reviewing her notes from yesterday's customer call. She flags an emotional cue the client gave—a moment of hesitation when discussing the budget. Instead of ignoring it, she builds it into her follow-up. At 9:00 a.m., her first meeting starts, a cross-functional internal call. Two colleagues disagree, and tension rises. Jen doesn't take sides. She reframes the disagreement, honors both views, and guides the team forward. That's Conflict Management and Teamwork in action.

At lunch, Jen journals. Quick bullet points: "Stressed in the morning—why?" She reflects. Adjusts. By 2:00 p.m., she's on a tough prospect call. The buyer challenges her on ROI. Jen doesn't get defensive. She breathes. Listens. She shifts from selling to understanding. The call doesn't end in a closed deal, but it ends with trust. That's EI. That's how the best reps operate.

Let's take a moment to go into the neuroscience of all this.

Your brain is wired to process emotion before logic. The limbic system—particularly the amygdala—fires up when you experience an emotional cue. It sends neurochemicals that influence memory, attention, and perception. If buyers feel safe, seen, and respected, they're more open to your message. Do they feel pressure, confusion, or threat? The prefrontal cortex (where rational decisions happen) goes offline. That's not theory—that's biology.

And that's why emotional regulation matters so much. When you manage your own nervous system, you help regulate theirs. A calm rep creates a calm buyer. That's how trust forms.

Now, what about sales leadership?

Coaching EI is one of the most powerful things you can do as a leader. But it requires intention. Start with modeling. When you make a mistake, own it. When a rep struggles, ask questions like "What's underneath that?" instead of jumping into solutions. Create psychological safety in your pipeline reviews. Recognize reps for emotional wins, not just closed revenue.

One of our clients, a medical device company, shifted its entire team culture when the leaders started coaching through the four EI domains.

They didn't just hit their number. They reduced turnover by 40 percent, increased NPS scores, and created an internal mentorship pipeline that doubled in a year. When EI is valued, performance follows.

In Summary:

Emotional intelligence is not a 'nice to have' in sales—it's the differentiator. From handling pressure and rejection to navigating internal collaboration and connecting deeply with clients, emotionally intelligent sales professionals outperform by operating with purpose, self-awareness, empathy, and resilience. In a world where every interaction matters, the 'real skills' sustain performance and build trust. Whether you are prepping for the biggest pitch of your career or rebounding from a hard-fought loss, emotional intelligence is your most powerful and underutilized advantage.

Takeaway Points

- Emotional intelligence is not a soft skill—it's a critical, measurable driver of sales performance that determines how you show up, connect, and close.

- Sales success isn't just about strategy and knowledge; it's about emotional regulation, connection, and adaptability across the entire deal cycle.

- Neuroscience confirms that emotion precedes logic in the brain's decision-making process—meaning the best sales professionals master emotional engagement first.

- Emotional competencies can be learned.

- Before each sales call, process in advance how you might emotionally respond within each step of the NeuroSelling® framework.

Three Practical Action Steps

1. Start a daily emotional journal. Reflect on emotional triggers, responses, and patterns influencing your tone and decision-making.
2. Before every sales call, use breathwork or visualization techniques to ground your mindset and activate emotional self-control.
3. After each interaction, review the four EI domains and ask yourself: What did I do well, and where can I improve across self-awareness, self-management, social awareness, and relationship management?

To Learn More About NeuroSelling®, go to
www.braintrustgrowth.com/neuroselling

PART III

THE DIGITAL EVOLUTION

THE AI REVOLUTION IN B2B SALES

"I'd rather see artificial intelligence than no intelligence."
—MICHAEL CRICHTON

The Convergence of Neuroscience and AI

THE FUSION OF NEUROSCIENCE and artificial intelligence represents perhaps the most significant transformation in sales since the digital revolution. As someone who has spent decades studying how the human brain makes decisions, I find it fascinating how AI is now being deployed to enhance—not replace—the very human elements of selling that we've explored throughout this book.

The question is no longer whether AI will impact B2B sales but how profoundly it already has and continues to do so. AI technologies are generating substantial value for global sales organizations, transforming how businesses approach customer relationships and decision-making processes. But beyond the impressive numbers lies a more nuanced story about how AI is transforming the profession while paradoxically making successful selling more human than ever.

To understand this apparent contradiction, we need to examine how AI intersects with the neuroscience principles that form the foundation of the NeuroSelling methodology.

How AI Is Transforming the Sales Landscape

During a recent keynote presentation I delivered to a group of sales executives, I asked how many were currently using AI tools in their sales process. Fewer than 20 percent raised their hands. When I asked how many expected to be using AI tools within the next two years, nearly every hand shot up.

This anecdote reflects what many industry observers are noting—AI adoption in sales is at an inflection point. The rapid growth is expected to continue as more organizations recognize the competitive advantages AI offers.

What's driving this rapid transformation? Three primary factors:

- Advancements in AI capability – Modern AI systems can now process unstructured data like emails, calls, and meeting transcripts to identify patterns invisible to human perception.

- Integration with existing workflows – New AI tools seamlessly integrate into CRM systems and communication platforms.

- Demonstrated ROI – Early adopters are showing measurable increases in productivity and revenue growth.

But the most significant change isn't technological—it's psychological. The most successful organizations are using AI not to automate human interaction but to enhance it. They're leveraging technologies that work similarly to how our brains naturally function, creating a powerful synergy between human connection and computational capability.

The Three Waves of AI in Sales

To understand where we are in the AI sales revolution, it helps to recognize the three distinct waves of AI adoption that have transformed the industry:

Wave 1: Automation (2015–2019) The initial wave focused primarily on automating repetitive tasks—email sequences, calendar scheduling, and basic prospecting. While valuable for efficiency, these tools had minimal impact on the actual sales conversation.

Wave 2: Augmentation (2020–2023) The second wave began enhancing the salesperson's capabilities with real-time insights, conversation analytics, and recommendation engines. These tools started influencing how salespeople communicated but largely operated as separate assistants.

Wave 3: Integration (2024–Present) We've now entered the third wave, where AI is becoming an integrated extension of the salesperson's capabilities—helping prepare for meetings, analyzing conversations in real-time, and proactively surfacing relevant information exactly when needed.

This third wave is particularly significant because it represents the true convergence of neuroscience and artificial intelligence. Today's most advanced sales AI doesn't just perform tasks faster—it mimics how the human brain processes information, identifies patterns, and makes decisions, but at a scale and speed no human could match.

AI-Enhanced NeuroSelling: The Five Key Applications

How does artificial intelligence specifically enhance the NeuroSelling methodology? Let's examine the five most impactful applications we're seeing with our clients today:

1. Personalization at Scale

One of the foundational principles of NeuroSelling is creating genuine personal connection. However, truly personalizing communication at scale has always been challenging.

AI is transforming this landscape through what we call "hyper-personalization." Systems now analyze thousands of data points about prospects—their communication style, industry challenges, digital behavior, and even psychological profiles—to help salespeople craft messages that resonate on a deeply personal level.

For example, one of our enterprise clients implemented an AI system that analyzes a prospect's LinkedIn posts, company announcements, and industry news to identify their specific goals, challenges, and communication preferences. The system then generates tailored messaging suggestions that align with the prospect's priorities, making the initial outreach significantly more relevant.

The results have been impressive, with substantial increases in response rates and meeting conversions. The critical insight: these weren't generic templates. The AI provided unique, prospect-specific insights that allowed salespeople to craft genuinely personalized messages addressing the prospect's actual priorities—not what the salesperson assumed they might be.

2. Emotional Intelligence Augmentation

Perhaps the most fascinating development is AI's growing ability to detect and respond to emotional signals. While AI cannot feel emotions, it can recognize them with remarkable accuracy through voice patterns, word choice, facial expressions, and digital behavior.

One promising application is real-time emotional intelligence coaching during virtual sales meetings. Using natural language processing and computer vision, these systems analyze a prospect's tone, facial expressions, and language patterns to detect engagement, confusion, interest, or skepticism—often picking up subtle cues the salesperson might miss.

The salesperson receives gentle nudges through a discreet earpiece or on-screen prompt: "They seem confused by that last explanation," or "They're showing strong interest when you discuss implementation timelines." This allows for real-time adjustments to the conversation based on the prospect's emotional response.

These systems aren't replacing the salesperson's emotional intelligence—they're enhancing it by providing additional data that would otherwise be missed, particularly in virtual settings where subtle emotional cues are harder to detect.

3. Narrative Optimization

Throughout this book, we've emphasized the power of strategic narratives in driving trust and creating urgency to change. AI is now helping salespeople craft more compelling narratives by analyzing what types of stories resonate most effectively with specific customer segments.

By analyzing thousands of successful sales conversations, AI systems can identify which narrative structures, examples, and framing devices create the strongest emotional impact for different buyer personas. This isn't about generating cookie-cutter scripts but providing evidence-based guidance on how to structure stories for maximum effect.

For example, one AI system we've worked with analyzes successful sales conversations to determine:

- Which problem framing creates the most urgency for different buyer roles

- What metaphors and analogies drive the clearest understanding

- Which customer examples create the strongest social proof for different industries

- How detailed the solution narrative should be based on the buyer's communication style

This helps salespeople structure their narratives more effectively while maintaining their authentic voice and tailoring to the specific customer context.

4. Objection Anticipation and Resolution

A particularly valuable application of AI is its ability to predict and prepare for potential objections before they arise. By analyzing patterns across thousands of similar sales conversations, AI can anticipate the specific concerns likely to emerge with a given prospect.

Financial services companies have found that their advisors who used AI objection anticipation systems increased their client conversion rates significantly. These systems don't just predict common objections—they provide specific guidance on how to address each concern based on the prospect's unique situation and communication preferences.

This application directly enhances the "Removing Barriers to Change" element of the NeuroSelling methodology. By preparing for objections before they arise, salespeople can address concerns proactively and maintain the conversation's momentum rather than being caught off-guard.

5. Continuous Improvement Loops

Perhaps the most powerful application is AI's ability to create personalized improvement loops for each salesperson. Traditional sales coaching has always been limited by sample size—a manager can only observe and provide feedback on a small percentage of a rep's conversations.

AI changes this equation dramatically. Systems can now analyze 100 percent of a salesperson's customer interactions—calls, emails, meetings—to identify specific patterns, strengths, and areas for improvement. This creates a continuous, data-driven feedback loop that accelerates skill development.

For example, one system might notice that a particular salesperson consistently loses momentum when transitioning from problem framing to solution presentation. The AI can provide specific guidance, relevant examples from top performers, and targeted practice opportunities to strengthen this specific skill.

The result is a highly personalized development path that adapts as the salesperson improves—focusing always on the highest leverage opportunities for growth rather than generic sales training.

The NeuroSeller's AI Toolkit

While the applications of AI in sales are expanding rapidly, several specific tools have emerged as particularly valuable for practitioners of the NeuroSelling methodology. Here are five essential categories that every modern sales professional should consider:

1. Conversation Intelligence Platforms

Platforms like Gong, Chorus, and ExecVision record, transcribe, and analyze sales conversations to identify patterns and coaching opportunities. The most advanced systems now provide real-time guidance during calls, highlighting when key topics are missed or suggesting effective responses to specific customer statements.

These platforms are especially valuable for analyzing how well salespeople execute the seven "P" narratives we've discussed throughout this book. They can identify whether the salesperson effectively establishes personal connection, positions the problem appropriately, and creates urgency to change.

2. Generative AI Research Assistants

Tools like Claude, ChatGPT, and Google's Bard have transformed how salespeople prepare for customer conversations. These systems can rapidly research industries, companies, and individual prospects to identify relevant insights, recent news, and potential pain points.

The key is using these tools strategically—not to generate generic outreach templates but to gather specific information that helps personalize your approach. For example, asking a generative AI to analyze a company's last three earnings calls to identify their strategic priorities and challenges mentioned by the leadership team.

3. Meeting Preparation and Follow-up Assistants

AI meeting assistants like Otter.ai, Fireflies, and Avoma not only transcribe conversations but also generate summaries, action items, and follow-up suggestions based on the content discussed. The most advanced systems create personalized follow-up recommendations that align with the specific concerns and interests expressed by the customer.

These tools help salespeople maintain momentum between conversations by ensuring nothing falls through the cracks, and every follow-up addresses the specific issues most important to the prospect.

4. Sentiment Analysis Tools

Platforms like Affectiva, Crystal, and Keen can analyze text communications, voice patterns, and even facial expressions (in video meetings) to detect emotional responses and communication preferences. These insights help salespeople adjust their approach in real time to better connect with prospects on an emotional level.

For example, sentiment analysis might reveal that a particular prospect responds more positively to concise, data-driven communication than to elaborate narratives, allowing the salesperson to adjust accordingly.

5. Predictive Analytics Platforms

Systems like People.ai, InsideSales, and Clari analyze historical data to predict which prospects are most likely to convert, what messaging will resonate most effectively, and which actions will advance the sale. These platforms help salespeople prioritize their time and efforts on the opportunities with the highest potential value.

The most sophisticated systems now incorporate neuroscience principles to predict how different types of buyers are likely to respond to specific messaging approaches, essentially codifying many of the NeuroSelling principles we've discussed throughout this book.

The Human-AI Partnership: Finding the Balance

Despite these remarkable advances, it's crucial to understand that AI is an enhancement to—not a replacement for—the fundamentally human process of sales. The most successful organizations maintain what we call a "human-in-the-loop" approach, where AI augments human capabilities rather than replacing them.

This balance is particularly important for NeuroSelling practitioners, who understand that genuine human connection remains the foundation of trust. Research is beginning to show that sales teams who delegate relationship-building aspects to AI see decreased performance compared to those who use AI primarily for research, analysis, and preparation.

The key is using AI for what it does best—processing vast amounts of information, identifying patterns, and generating insights—while reserving uniquely human skills like empathy, ethical judgment, and creative problem-solving for the salesperson.

For NeuroSelling practitioners, this means using AI to enhance your understanding of the customer's world, optimize your narratives, and identify the most effective approach—while maintaining authentic human connection throughout the process.

Ethical Considerations in AI-Enhanced Selling

As we embrace the potential of AI in sales, we must also grapple with important ethical questions. The power to influence buying decisions comes with significant responsibility, and several ethical considerations deserve particular attention:

Transparency and Trust

Prospects deserve to know when AI is being used in the sales process, particularly when it involves analyzing their data or behavior. Organizations that maintain transparency about their AI usage actually build greater trust with customers.

The MIT Sloan Management Review has published research indicating that B2B buyers respond positively when informed that AI is being used to personalize their experience but react negatively when they discover AI involvement that hasn't been disclosed.[1]

Privacy and Data Usage

The effectiveness of sales AI depends on the data it can access and analyze. Organizations must be meticulous about compliance with privacy regulations and transparent about how customer data is used.

Beyond legal requirements, ethical AI usage requires thoughtful consideration of what data analysis is appropriate and beneficial versus what might feel invasive to prospects. The line between helpful personalization and uncomfortable surveillance can be thin.

Algorithmic Bias

AI systems learn from historical data, which means they can perpetuate or even amplify existing biases in sales approaches. For example, if historical data shows that salespeople have primarily targeted specific demographics, AI recommendation systems might continue to prioritize those same groups while overlooking valuable but underrepresented prospects.

Organizations must actively monitor for bias in their AI systems and implement safeguards to ensure fair and inclusive sales practices.

Human Autonomy and Skill Development

As AI becomes more sophisticated, there's a risk that salespeople might over-rely on its recommendations, potentially diminishing their own skills and judgment over time. Organizations should design AI systems that enhance human capabilities rather than replace human decision-making entirely.

The most effective approach treats AI as a coach and advisor rather than an autopilot, ensuring that salespeople maintain and develop their core skills while benefiting from AI insights.

Preparing for the AI-Enhanced Future

The integration of AI into B2B sales is accelerating rapidly, and organizations that adapt effectively will have a significant competitive advantage. Here are five key recommendations for salespeople and sales leaders preparing for this future:

1. Invest in AI Literacy

Every sales professional needs a baseline understanding of AI capabilities, limitations, and applications. This doesn't mean becoming a data scientist but rather developing sufficient knowledge to effectively collaborate with AI systems and critically evaluate their recommendations.

2. Double Down on Distinctly Human Skills

As AI handles more analytical and administrative aspects of selling, distinctly human capabilities become even more valuable. These include emotional intelligence, ethical judgment, creativity, and the ability to build genuine trust-based relationships.

3. Adopt a Test-and-Learn Approach

The AI landscape is evolving rapidly, and no one has all the answers yet. The most successful organizations maintain a culture of experimentation—testing new AI applications, measuring results, and continuously refining their approach.

4. Redesign Sales Processes with AI in Mind

Rather than simply layering AI onto existing processes, forward-thinking organizations are redesigning their entire sales approach to leverage the unique capabilities of both humans and AI. This often means rethinking role definitions, compensation structures, and performance metrics.

5. Develop an Ethical Framework

Every organization should establish clear guidelines for ethical AI use in sales, addressing questions of transparency, privacy, bias, and human autonomy. These guidelines should evolve as AI capabilities advance and new ethical considerations emerge.

The Future of AI in NeuroSelling

As we look toward the horizon, several emerging developments promise to further transform how AI enhances the NeuroSelling methodology:

Multimodal Emotional Analysis

Next-generation systems will analyze not just language and tone but also facial expressions, body language, and even physiological signals (through wearable devices) to provide a comprehensive understanding of a prospect's emotional state. This will allow for even more precise calibration of sales approaches based on the prospect's receptivity.

Augmented Reality Sales Environments

AR technologies will create immersive presentation environments where salespeople can visualize complex solutions and their impact in ways that create powerful emotional engagement. AI will optimize these experiences in real time based on the prospect's responses.

Neural-Symbolic Integration

Emerging AI approaches combine the pattern recognition capabilities of neural networks with symbolic reasoning systems that understand causal relationships. This hybrid approach will enable AI to not just recognize patterns in sales data but also understand why certain approaches work in specific contexts.

Decentralized AI Assistants

Rather than relying on centralized AI platforms, salespeople will increasingly work with personalized AI assistants that learn their specific communication style, strengths, and areas for development. These assistants will provide increasingly customized guidance tailored to each salesperson's unique development needs.

Conclusion: The Augmented NeuroSeller

The integration of AI into sales isn't about replacing the human element—it's about amplifying it. By handling routine analysis and providing data-driven insights, AI frees salespeople to focus on what they do best: building genuine connections, understanding complex human needs, and creating transformative value for customers.

For practitioners of the NeuroSelling methodology, AI represents a powerful accelerant. It helps identify which aspects of the approach are working most effectively for specific customer segments, provides real-time guidance on narrative optimization, and creates personalized development paths to strengthen each salesperson's capabilities.

The most successful salespeople in this new landscape won't be those who resist AI or those who rely on it entirely. They'll be the "augmented NeuroSellers" who leverage AI's analytical power while maintaining the authentic human connection at the heart of all meaningful sales relationships.

As we navigate this transformation, one thing remains constant: the fundamental neuroscience principles that drive human decision-making. AI may change how we apply these principles, but the underlying science of how people build trust, process information, and make decisions remains the foundation of effective selling.

The future belongs to those who embrace both the science of human decision-making and the technological tools that help us apply that science more effectively than ever before.

NEUROSELLING IN A DIGITAL WORLD

"Technology is nothing. What's important is that you have a faith in people, that they're basically good and smart, and if you give them tools, they'll do wonderful things with them."

—STEVE JOBS

The New Digital Reality

The world of selling has undergone a seismic shift. When I wrote the first edition of this book, we were just beginning to understand the impact of digital technologies on the sales process. Today, digital selling isn't just an option—it's a fundamental requirement.

The COVID-19 pandemic accelerated digital transformation by several years, according to industry research.[1] Virtual selling, once considered a supplementary channel, has become a primary mode of customer engagement. Recent studies indicate that a vast majority of B2B sales interactions between suppliers and buyers will occur in digital channels within the next few years.[2]

What does this mean for NeuroSelling practitioners? Do the neuroscience principles we've explored throughout this book still apply in virtual environments? The answer is a resounding yes—but with important nuances and adaptations.

The fundamentals of human decision-making haven't changed. Your customers' brains still process information from the inside out, starting with emotion and trust before engaging in rational evaluation. They still have the same barriers to change and cognitive biases. However, the digital medium introduces unique challenges in how we create connection, build trust, and drive urgency to change.

In this chapter, we'll explore how to apply NeuroSelling principles in digital selling environments—from video calls and virtual presentations to digital content and social selling. You'll learn specific strategies to overcome the constraints of digital channels while leveraging their unique advantages.

The Neuroscience of Digital Interaction

Research on the neuroscience of digital interaction reveals fascinating insights about how our brains process virtual communication differently from in-person exchanges. These differences have profound implications for how we approach digital selling.

The Digital Trust Barrier

Research has shown that establishing trust through digital channels takes significantly longer than in face-to-face settings.[3] Why? Because our brain's limbic system—responsible for emotional connection and trust—evolved to process the full spectrum of human interaction cues.

In virtual environments, we lose access to many subtle trust signals:

- Microexpressions – Fleeting facial expressions lasting fractions of a second that reveal genuine emotions
- Body language – Posture, gestures, and movement that signal openness and trustworthiness

- Environmental context – Shared physical space that creates psychological proximity
- Scent and pheromones – Unconscious chemical signals that influence trust responses
- Touch – Physical connection that triggers oxytocin release (the "trust hormone")

This doesn't mean trust is impossible in digital environments—it simply means we must be more intentional about creating it. Successful digital NeuroSellers compensate for these missing signals by amplifying other trust-building elements and leveraging the unique capabilities of digital channels.

Zoom Fatigue and Cognitive Processing

Neuroscientists have revealed that digital interactions impose a higher cognitive load than in-person communication. When we interact through screens, our brains work harder to:

- Process nonverbal cues that appear distorted or delayed
- Maintain constant gaze (unlike natural eye contact patterns)
- Monitor our own appearance on screen
- Filter out distractions in our physical environment
- Compensate for audio delays and technical issues

This increased cognitive effort, now commonly known as "Zoom fatigue," has significant implications for sales conversations. When your customer's brain is allocating extra resources to these processing tasks, they have less cognitive capacity available for evaluating your message and making decisions.

The practical consequence? Digital sales conversations need to be more focused, more engaging, and more cognitively efficient than their in-person counterparts.

The Digital Empathy Gap

Perhaps most concerning for sales professionals is what researchers call the "digital empathy gap." Brain imaging studies have shown that the same message delivered digitally versus in person activates empathy circuits in the brain less strongly. This empathy reduction makes it harder for customers to:

- Feel your authentic concern for their challenges
- Experience emotional engagement with your narratives
- Envision themselves benefiting from your solution
- Connect with the urgency of solving their problems

This empathy gap is especially problematic for NeuroSelling, which relies heavily on emotional connection and narrative engagement to drive change. The good news is that specific digital communication strategies can significantly narrow this gap.

Mastering the NeuroSelling Narratives in Digital Environments

Let's explore how to adapt each of the NeuroSelling narratives for maximum impact in digital environments:

1. Personal Story (Digitally Enhanced)

Your personal "Why" Story becomes even more critical in digital environments where trust is harder to establish. However, it needs adaptation for digital delivery.

Digital Enhancements:

- Visual reinforcement – Have one to two simple personal photos ready to share at key moments (but don't overdo it with a slideshow).
- Increased expressiveness – Use more vocal variety and facial animation than would feel natural in person.
- Direct eye contact – Look at your camera (not your screen) during the most emotionally significant moments.

- Setting context – Briefly acknowledge the digital medium: "Even though we're meeting virtually today, I'd like to share something personal about why I do this work."

Example: One of our clients, a medical device sales leader, begins her virtual meetings by showing a single photo of her grandfather, who received a pacemaker that extended his life by fifteen years. She shares her story while maintaining direct eye contact with the camera, using an expanded vocal range to convey emotion. She concludes by asking the physician about their "why" while returning to standard eye engagement.

2. Prospect Story (Digitally Enhanced)

Demonstrating that you understand your prospect's world becomes more challenging—but also more differentiating—in digital environments. Research shows that personalization has an even stronger impact in digital interactions.[4]

Digital Enhancements:

- Pre-meeting research – Use digital tools to gather more specific insights about your prospect before meetings.
- Visual mapping – Create simple visual representations of your prospect's goals and challenges to share on-screen.
- Digital empathy statements – Acknowledge the unique challenges of their current work environment.
- Shared digital whiteboarding – Use collaborative tools to map out their situation together.

Example: A financial services advisor we work with creates a simple "day-in-the-life" visualization for each prospect before their meeting. During the call, she shares her screen and walks through this visual representation, asking the prospect to help her refine it. This collaborative process demonstrates deep understanding while creating engagement through active participation.

3. Problem Story (Digitally Enhanced)

Problem framing requires particular attention in digital environments because of the reduced emotional impact. Research shows that risk perception is notably lower in digital versus in-person communications.[5]

Digital Enhancements:

- Visual problem quantification – Create clear, simple visuals showing the cost of the problem.
- Concrete examples – Use more specific, vivid examples than you would in person
- Emotional bridging – Explicitly connect problems to emotions: "Many clients tell me this situation keeps them up at night."
- Digital silence – After presenting a significant problem, pause longer than feels comfortable to allow processing.

Example: An enterprise software salesperson uses a simple animated graphic showing how a $5,000 monthly efficiency problem compounds to $300,000 over five years. He then uses the "digital silence" technique—letting the graphic remain on screen for a full five to seven seconds without speaking—creating space for the emotional impact to register.

4. Product/Solution Story (Digitally Enhanced)

Solution presentations often suffer the most in digital environments because they typically involve complex information that's harder to process through screens. The key is simplification and engagement.

Digital Enhancements:

- Progressive disclosure – Reveal information in smaller, more digestible chunks.
- Interactive demonstrations – Use screen sharing for live demonstrations rather than passive presentations.

- Visual metaphors – Use simple visual metaphors that connect your solution to familiar concepts.
- Participation prompts – Create moments for active engagement every three to five minutes.

Example: Instead of a typical product demo, a SaaS company we work with uses a "choose your own adventure" approach. They prepare multiple demonstration paths and let the customer direct which capabilities to explore first based on their priorities. This maintains engagement while creating a sense of control.

5. Proof Story (Digitally Enhanced)

Validation becomes particularly important in digital environments because risk perception is heightened. Research shows that specific types of proof have dramatically different impact in digital settings.[6]

Digital Enhancements:

- Video testimonials – Short video clips of actual customers have much greater impact than quoted testimonials.
- Relevant specificity – Focus on fewer, more directly relevant examples with greater detail.
- Data visualization – Use simple, clear visualizations of results rather than numbers alone.
- Social proximity – Emphasize similarities between your prospect and reference customers.

Example: A manufacturing equipment provider creates sixty-second video testimonials from customers in the same industry as their prospect. Rather than generic success stories, these videos address specific concerns the prospect has raised during the conversation. They introduce these with: "I thought you might appreciate hearing directly from someone who had the exact same concern about . . ."

Digital Trust Accelerators

While digital environments present trust challenges, they also offer unique opportunities to accelerate trust formation when used strategically. Here are five digital trust accelerators that leverage the unique advantages of virtual selling:

1. Micro-Demonstrations of Competence

In digital settings, small demonstrations of technical competence create disproportionate trust impact. When you demonstrate mastery of the digital medium itself, customers subconsciously extend that perception of competence to your broader expertise.

Implementation:

- Master your virtual meeting platform's features.
- Have backup plans for technical issues and execute them smoothly if needed.
- Provide meaningful pre-meeting materials that demonstrate value.
- Follow through instantly on digital commitments ("I'll send that right now").

2. Digital Reciprocity Loops

The principle of reciprocity—that people feel obligated to give back when they receive something—works even more powerfully in digital environments when implemented correctly.

Implementation:

- Provide unexpected valuable insights before asking for anything.
- Create custom research or analysis specific to the prospect's situation.
- Use digital tools to deliver value between formal meetings.
- Make reciprocity visual rather than just verbal.

3. Digital Consistency Anchors

In virtual selling environments where interactions may be more sporadic, demonstrating consistency becomes a powerful trust mechanism.

Implementation:

- Create consistent visual and verbal patterns across all digital touchpoints.
- Establish and maintain reliable communication rhythms.
- Use progress-tracking visuals to demonstrate momentum.
- Reference and build upon previous conversations explicitly.

4. Virtual Background Psychology

Your digital environment communicates powerful subconscious messages about your credibility and trustworthiness. Research shows that carefully chosen backgrounds can influence perception significantly.

Implementation:

- Choose environments that balance professionalism with authenticity.
- Consider what's visible in your physical or virtual background.
- Ensure lighting highlights your face clearly (especially your eyes).
- Remove visual distractions that compete for attention.

5. Digital Immediacy

The ability to provide immediate responses and resources creates a powerful trust advantage unique to digital selling.

Implementation:

- Prepare resources in advance to share instantly during calls.
- Use collaborative tools for real-time problem-solving.
- Follow up immediately after meetings with promised resources.
- Leverage automation for consistent, timely communications.

Managing the Digital Change Resistance

The barriers to change we explored earlier in the book manifest differently in digital environments. Here's how to identify and address these digital-specific resistance patterns:

Digital Anxiety Amplification

In virtual settings, anxiety about change can be amplified by the lack of reassuring physical presence. This manifests as prolonged silence, reduced engagement, or sudden shifts to procedural questions.

Counterstrategies:

- Acknowledge the uncertainty explicitly but confidently.
- Use collaborative visualization tools to make change concrete.
- Provide more frequent reassurance and validation.
- Create digital "change road maps" with clear, manageable steps.

Digital Isolation Effect

The fear of being alone in making a change becomes more pronounced in digital environments where physical separation reinforces psychological isolation.

Counterstrategies:

- Create virtual connections to existing customers.
- Use video testimonials from similar customers.
- Establish digital user communities for peer support.
- Demonstrate broad adoption through visual data.

Digital Status Quo Anchoring

The distance created by digital interaction often strengthens attachment to the known and familiar—what psychologists call "status quo bias."

Counterstrategies:

- Create visual contrast between current state and future state.
- Use digital simulation to make the future state feel more concrete.
- Implement small, low-risk digital pilots before full commitment.
- Create digital "before and after" scenarios.

Digital Decision Fatigue

The cognitive load of virtual meetings accelerates decision fatigue, making change decisions particularly challenging near the end of digital interactions.

Counterstrategies:

- Schedule decision conversations early in the day.
- Break complex decisions into smaller digital commitments.
- Create visual decision frameworks that simplify choices.
- Provide clear, simple next steps that feel manageable.

Digital Question Mechanics

The art of asking effective questions takes on new dimensions in digital environments. Research reveals that question patterns that work well in person often fail in virtual settings.

Digital Question Principles

- Shorter question strings – Limit yourself to one to two questions at a time versus three to four in person.
- Increased processing time – Provide 30 percent more wait time after questions.
- Visual reinforcement – Display important questions on screen briefly.
- Progressive specificity – Begin with broader questions, then narrow focus.

- Engagement validation – Use visual and verbal cues to confirm understanding.

Digital Question Types with High Impact

Certain question types perform exceptionally well in digital environments:

- Contrast questions show particularly strong digital engagement: "How does your current approach compare to what you'd ideally like to be doing?"
- Visualization questions create powerful mental engagement: "If we could solve this problem completely, what would that make possible for you?"
- Reflection questions cut through digital distraction: "Looking at these three challenges we've discussed, which one keeps you awake at night?"
- Prioritization questions create clarity in complex digital discussions: "If you could only solve one of these issues immediately, which would create the most value?"

The Digital-Physical Integration Strategy

While this chapter focuses on digital selling, the most effective modern sales approach integrates digital and physical interactions strategically. Research shows that hybrid selling approaches—combining digital and in-person touchpoints—outperform either approach alone by a significant margin.

The Optimal Hybrid Selling Sequence

Based on neuroscience research, certain interactions are more effective in specific channels.

Digital Optimal:

- Initial discovery conversations
- Regular progress updates
- Technical demonstrations

- Document reviews and collaboration
- Quick decision confirmations
- In-Person Optimal:
- Deep trust establishment
- Complex problem exploration
- High-stakes negotiations
- Multi-stakeholder consensus building
- Implementation kickoffs

The key is designing your sales process to leverage each channel for what it does best rather than treating digital as a compromise.

Cross-Channel Consistency

As you move between digital and physical environments, maintaining neurological consistency becomes essential. This means creating recognizable patterns across channels that reinforce trust and momentum:

- Visual consistency – Use the same visual frameworks and metaphors
- Narrative continuity – Reference and build upon previous conversations
- Relationship memory – Acknowledge personal details across channels
- Incremental commitment – Build upon previous agreements

Creating Neurocentric Digital Content

Beyond direct interactions, your digital content—from emails and social posts to videos and website pages—can either support or undermine your NeuroSelling approach. The most effective digital content follows neurocentric design principles:

The Neuroscience of Digital Content Engagement

Research has identified specific content characteristics that activate deeper brain engagement in digital environments:

- Personal relevance – Content that feels specifically created for the recipient
- Narrative structure – Information organized as a coherent story
- Visual processing ease – Content designed for low cognitive processing effort
- Emotional anchoring – Content that connects to core emotional drivers
- Action clarity – Clear, simple next steps without decision complexity

Applying NeuroSelling to Digital Content

Your digital content should follow the same inside-out approach as your direct conversations:

Personal Connection First:
- Begin with empathy for the reader's situation
- Use conversational, authentic language
- Share relevant personal insights where appropriate
- Focus on the reader, not on your company

Problem Before Solution:
- Address specific challenges before introducing offerings
- Quantify problems in meaningful terms
- Create emotional engagement with the problem
- Use visual problem framing when possible

Simplicity and Focus:
- Address one primary concept per content piece
- Use visual hierarchy to guide attention

- Eliminate unnecessary complexity
- Create white space for mental processing

Direct Relevance:
- Customize content for specific industries or roles
- Reference the recipient's specific situation when possible
- Provide context for why this matters to them specifically
- Connect to their business objectives

The Future of Digital NeuroSelling

As we look to the horizon, several emerging technologies promise to further transform how we apply NeuroSelling principles in digital environments:

Augmented Reality Sales Experiences

AR technologies are creating immersive presentation environments that dramatically increase emotional engagement and memory retention. Early research shows that AR product demonstrations increase purchase intent significantly compared to traditional digital presentations.

These technologies allow customers to:

- Experience products in their actual environment
- Interact with virtual demonstrations physically
- Visualize complex concepts spatially
- Create stronger emotional connections with solutions

Emotional AI Integration

Advanced emotional analysis technologies are beginning to bridge the empathy gap in digital selling. These systems analyze facial expressions, voice patterns, text sentiment, and engagement signals to provide real-time guidance on emotional connection.

The most promising applications include:

- Real-time emotion recognition during virtual meetings
- Message optimization based on emotional response patterns
- Personalized content sequencing based on engagement data
- Coaching on emotional intelligence factors in digital communication

Immersive Collaborative Environments

Beyond traditional video conferencing, new immersive collaboration platforms are creating shared digital spaces that more closely replicate the neurological benefits of in-person interaction.

These environments offer:

- Spatial audio that mirrors natural conversation dynamics
- Avatar-based nonverbal communication
- Shared digital objects for collaborative manipulation
- Persistent digital spaces that create continuity between interactions

Conclusion: The Digitally Enhanced NeuroSeller

The principles of NeuroSelling remain constant whether you're sitting across the table from a prospect or connecting through screens. The human brain still processes information from the inside out, starting with emotion before engaging logic. Trust still forms the foundation of influence. Narratives still drive engagement and memory.

What changes in digital environments is not the underlying neuroscience but how we apply it. The constraints of digital channels require more intentional trust-building, more engaging narrative delivery, and more active management of cognitive load. The opportunities of digital selling—from enhanced personalization to immersive experiences—offer powerful new ways to create connection and drive change.

The most successful sales professionals in the coming years won't be those who resist digital transformation or those who abandon human connection in favor of automation. They'll be the digitally enhanced NeuroSellers who leverage technology to amplify their human capabilities—creating deeper connections, delivering more impactful narratives, and driving more meaningful change for their customers.

CASE STUDIES

CASE STUDY 1: Transforming a Clean Energy Innovator

The Challenge

The client, a clean energy storage and power conversion company that designs, builds, and executes systems to dramatically advance the clean energy future, approached us with several strategic challenges they needed to overcome to meet their ambitious growth targets. The company's leadership recognized that despite having exceptional technology and engineering capabilities, they needed to transform their approach to market positioning, sales enablement, and leadership development.

They were looking for a partner to help navigate several strategic challenges as part of their growth agenda. First, they wanted to develop and launch new branding, positioning, direction, and content marketing strategies. In parallel, they aimed to expand their influence within the clean energy sector by growing their thought leadership through content marketing.

Another critical area of focus was establishing a robust sales enablement foundation, which was essential for effective execution and growth in their competitive markets. Furthermore, they were dedicated to creating a culture of world-class coaches, which necessitated designing and building a sustainable training approach to support continuous development and excellence.

The NeuroSelling® Solution

Our partnership with the client began with a marketing consulting agreement but quickly evolved into a comprehensive transformation initiative.

We played a pivotal role in the client's new company branding and positioning, providing comprehensive digital go-to-market consulting. We conducted a detailed competitive analysis and generated targeted content through our NeuroMarketing™ services. Using NeuroMessaging™, we developed storyboards for two key verticals and delivered NeuroSelling® skills training focused on the clean energy and power systems sectors.

Additionally, we offered coaching reinforcement sessions, NeuroCoaching™ for the leadership team, executive coaching, and facilitated the development of their Digital Academy.

As their VP of Sales & Marketing noted, "Their attention to detail, structured programs, project management, commitment to the partnership, and execution of deliverables has been second to none. As a result, we progressed from what was initially a marketing consulting agreement to a full-blown partnership, including NeuroMessaging, NeuroSelling, and NeuroCoaching for the second half of the year. The partnership has been so successful that we plan to extend the relationship."

The Results

The impact of implementing NeuroSelling principles and comprehensive transformation was significant. The client saw their bookings grow from $43.6 million in 2020 to $88.7 million in 2021.

This surge in bookings was anticipated to transition into a backlog and an increase in EBITDA by 2022. The upward trajectory culminated in the sale of the company in 2022, marking a pivotal milestone in its corporate journey.

Key Takeaways

This case demonstrates how the integration of NeuroSelling with complementary approaches to marketing and leadership development can transform an organization's performance. For this clean energy client, the key elements that drove success were the following:

1. Developing messaging that resonated at a deeper level with their target audience
2. Equipping the sales team with sector-specific NeuroSelling skills
3. Building a coaching culture that reinforced new approaches and ensured continuous improvement
4. Creating a consistent, integrated approach across branding, marketing, sales, and leadership

The result was not just improved sales performance but a transformation that significantly increased the company's value and positioned it for a successful exit.

CASE STUDY 2: Transforming a $3 Billion Enterprise Sales Team

The Challenge

The client, a leading provider of human capital management solutions offering services in human resources, payroll, employee benefits, and insurance, faced significant challenges with their sales approach. In fiscal year 2024, they reported $5.3 billion in revenue and $2.17 billion in operating income, but their sales teams were struggling with several key issues.

Sales representatives were conducting product-focused, transactional conversations that led to little differentiation in the marketplace. This approach made prospects view their solutions as commodities, resulting in frequent discount requests and price negotiations.

Additionally, the sales team lacked a profound understanding of their prospects' goals and challenges, which led to ineffective communication and difficulty establishing value. Prospects showed little urgency to buy, often fearing change and seeing it as unnecessary.

These issues manifested in measurable business problems: low closing ratios of just 18 percent and flat to declining revenue per sales representative and per deal.

The NeuroSelling® Solution

We implemented our comprehensive NeuroSelling program, which fundamentally changed how the client's sales teams approached customer conversations.

The program trained sales teams on establishing trust, connection, and credibility with prospects. It taught representatives about the science of the "buying brain" and how to create compelling, value-focused messaging that resonated with decision-makers.

A key component of the training included visual storytelling techniques and methods for effectively contrasting price with value. The approach shifted the sales team's mindset from selling products to facilitating customer choices, with a much stronger focus on addressing customer goals and challenges.

One significant benefit emerged for sales leadership: Sales managers found the new methodology easier to coach, and continuous support systems helped reinforce skills development over time.

The Results

The impact of the NeuroSelling program was substantial and measurable:

- After completing the training, 108 out of 135 sales representatives increased their total revenue by an average of $5,702 per month. This improvement totaled an additional $7,389,792 in revenue over a twelve-month period.
- Revenue per deal increased by $428 for 60 percent of the representatives, demonstrating their enhanced ability to communicate value and reduce discount requests.
- Perhaps most impressively, the closing ratio improved dramatically from 18 percent to 35 percent, significantly exceeding the industry standard of 20–25 percent.

As one sales representative noted, "I couldn't believe how different my sales calls were after going through this program. It was night and day!"

The program's effectiveness was further validated when the client decided to make a deeper commitment: "Based on results, we have not only adopted this program but licensed the rights to make it our own!"

Key Takeaways

This case illustrates how the NeuroSelling methodology can transform sales performance by addressing fundamental aspects of buyer psychology:

1. The shift from product-focused to customer-focused conversations creates meaningful differentiation in the marketplace.
2. Teaching sales teams to understand and address customer goals and challenges improves communication effectiveness.
3. Visual storytelling and value-focused messaging techniques help overcome buyer resistance and price sensitivity.
4. A consistent, coachable methodology provides long-term sustainability and continuing performance improvement.

As one executive from the client company summarized, "The world had changed . . . to compete, we needed a better way to tell a better story. This was it."

CASE STUDY 3: Transforming a Mid-Market Specialty Services Company

The Challenge

The client, an underground utility/safety company specializing in the detection of underground utilities, video pipe inspection, and the scanning of concrete structures, faced significant challenges in both their marketing and sales operations.

On the marketing side, they struggled with a relatively underdeveloped digital marketing approach. Their limited expertise in content marketing and paid ad management was costing them tens of thousands of dollars per month unnecessarily, with suboptimal results to show for this significant investment.

The sales challenges were equally pressing. The majority of their sales team had originated from the project delivery side of the business and had no formal sales training or coaching. This technical background meant most representatives were comfortable discussing the technical aspects of their services but struggled with effective sales communication. The company found it difficult to transform their team from reactive, transactional communicators to strategic, solution-focused sales professionals.

These challenges were limiting the company's growth potential despite having excellent technical capabilities and services.

The NeuroSelling® Solution

We implemented our comprehensive "Velocity" growth program across the entire organization, addressing both the marketing and sales challenges simultaneously.

For the marketing team, we leveraged five marketing resources to handle all aspects of content creation, social media management, and the overall execution of the paid ads strategy. This brought professional marketing expertise to a team that had previously lacked specialized digital marketing knowledge.

On the sales side, we delivered multiple integrated solutions:

- Created NeuroMessaging™ storyboards that helped representatives communicate value more effectively
- Delivered NeuroSelling® skills training to outside and inside sales teams, as well as marketing and operations departments
- Provided reinforcement coaching to all NeuroSelling® participants to ensure the new approaches became embedded in daily practices
- Implemented NeuroCoaching™ for the leadership team
- Offered strategic and executive coaching for leadership
- Created and supported the development of their Digital Academy to institutionalize the new approaches

This all-encompassing strategy made sure that marketing was producing high-quality leads and that the sales team had the tools necessary to successfully convert those leads using neuroscience-based communication techniques.

The Results

The impact of our partnership was substantial and measurable:

- On the sales/revenue side, bookings grew from $76 million to $101 million over a twelve-month period, representing a 33 percent sales increase. This growth culminated in a record month of more than $10 million in sales established in August 2021.
- The marketing improvements were equally impressive. Initial ROI metrics met or exceeded lead quality and volume targets while achieving a 40 percent reduction in monthly ad spend, saving approximately $20,000 per month. The optimized ads strategy generated an average of $50,000 per day in sales revenue.

As the CEO noted: "If you are looking for ELITE growth partners for your company, look no further than the team at Braintrust. I was fortunate enough to be introduced to them about fifteen months ago, and this team has made an incredible impact on our sales, marketing and operations teams."

Key Takeaways

This case study demonstrates several important principles about the NeuroSelling® approach:

1. Integrating marketing and sales transformations creates powerful synergy that accelerates revenue growth.
2. Technical experts can become effective sales professionals when given the right training and coaching.
3. A neuroscience-based approach to sales communication helps transform transactional conversations into strategic, solution-focused interactions.
4. Professional marketing expertise can simultaneously increase lead quality while reducing ad spend.
5. A comprehensive program that addresses both skill development and leadership coaching creates sustainable change.

The client's success in achieving a 33 percent revenue increase while simultaneously reducing marketing costs highlights the transformative potential of the integrated NeuroSelling® approach.

CASE STUDY 4: Revitalizing a Global Pharmaceutical Company

The Challenge

The client, a global pharmaceutical company specializing in brain diseases, faced dual challenges in their sales approach and organizational culture. With approximately 5,400 employees operating in more than 100 countries and reaching over 8 million people daily, they were a significant player in their field. Focusing exclusively on psychiatric and neurological disorders, the client maintained R&D centers in Denmark and the US, along with production facilities in France, Italy, and Denmark.

Despite their impressive global footprint, they struggled with two critical issues:

- Sales Challenges: Their sales strategy relied heavily on data and product features, an approach that increasingly failed to engage healthcare professionals effectively. This product-centric approach resulted in modest growth as it did not build strong customer relationships or address customer needs in a meaningful way. A mature brand with plateauing sales reflected this challenge.

- Organizational Challenges: High employee turnover (25 percent) had resulted in one-third of the organization having less than two years of tenure. This significant "churn" disrupted team cohesion, hindered the development of a positive culture, and limited promotional opportunities, leading to a disengaged workforce.

The NeuroSelling® and NeuroCoaching® Solution

Our partnership with the client involved a comprehensive, multi-year approach addressing both their sales methodology and leadership development:

Sales Transformation: We implemented our NeuroSelling® methodology, which drove a fundamental shift in their approach. Rather than focusing primarily on data and product features, we helped their teams focus on customer outcomes, develop advanced storytelling techniques, and adopt a consultative problem-solving approach. This transformed customer interactions from transactional to consultative, enabling sales representatives to connect more effectively with healthcare professionals.

Cultural Transformation: Through our NeuroCoaching™ program, we helped transform the client's culture by translating neuroscience principles into a practical model for leadership development. This created positive momentum throughout the organization and significantly increased internal promotional opportunities. The coaching initiative helped crystallize a foundational vision for their business unit, each leader, and their employees.

As their vice president of sales noted: "NeuroCoaching helped my team translate neuroscience principles into a practical model that will have a sustaining impact on team member development and performance. We were looking for a sustainable coaching model, and we got it."

The Results

The impact of this dual transformation was substantial:

- Over a three-year partnership, revenue increased from $1 billion to $1.15 billion, representing growth of $150 million.

- They successfully grew a mature brand with plateauing sales from flat to an increase of nearly $200 million in year-over-year sales.

- Perhaps most significantly, employee turnover decreased dramatically from 25 percent to 8 percent, reflecting a much healthier organizational culture and increased employee engagement.

- The client's leadership recognized the value of this transformation: "As a team, we crystallized a foundational vision for our business unit, each leader, and our employees. The program will take us from a good company to a great leadership organization."

Key Takeaways

This case study demonstrates the power of integrating NeuroSelling® with NeuroCoaching™ to address both external sales effectiveness and internal organizational health:

1. Even in highly scientific fields like pharmaceuticals, moving beyond data-driven selling to outcome-focused conversations drives significant revenue growth.

2. Advanced storytelling techniques can revitalize mature products with plateauing sales.

3. Applying neuroscience principles to coaching creates sustainable leadership development models.

4. Reducing employee turnover has measurable financial benefits in addition to cultural improvements.

5. A comprehensive approach addressing both sales methodology and leadership development delivers superior results compared to focusing on either in isolation.

The three-year ongoing partnership demonstrates the sustained impact of these approaches when they become embedded in an organization's culture and operations.

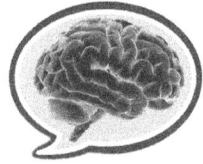

APPENDIX: NEUROSELLING JUMP-START

IT'S TAKEN ME a lifetime to put the pieces of NeuroSelling together. I hope it saves you years of learning, helps increase your sales dramatically, and enables you to help far more customers. Most importantly, I hope it helps you be a more impactful communicator in every area of your life. However, just like your prospects, you have to choose to implement the information contained within these pages.

I understand this book contains a lot of information. It can be hard to absorb it all in one sitting. In some ways, it can feel like I've held a firehose a few feet from your face and pulled the lever. Even our clients who attend our full workshops feel a bit overwhelmed at first. After all, I am asking you not only to leave your existing safety box but also giving you a scientific rationale as to why. Your fear of change, coupled with your potential fear of not being able to execute each piece of this approach, can be a challenge. My guess is that many of you recognize areas covered in this book as concepts you've been doing well but possibly didn't know the reason why they worked. For other areas, maybe the concepts here help reveal places in your customer conversation that can still improve.

As a way to "bring it all together" in one place, you may find the following road map or "cheat sheet" helpful. After reading all this, you may be looking for a "cheat sheet" to help you get started. I've got you covered!

The NeuroSelling Implementation Framework

#1. People buy from people they trust.

There are two types of trust, personal and professional. Not until your prospect trusts you personally will they drop their self-preservation shield in order to see how you may help them professionally. Feelings of trustworthiness drive up oxytocin, the "trust" neurochemical. The quickest way to do this is with the first of your "P" narratives, your personal "why" story.

Action Steps:
- Draft your personal "Why" Story using the structure outlined in Chapter Nine.
- Practice delivering it in under two minutes.
- Record yourself telling it and review for authenticity and impact.
- Test different versions with colleagues for feedback.

#2. Your prospect needs to see that you understand their world.

Instead of asking a hundred "probing" questions, present them with three to five typical goals or objectives someone in their role tends to focus on by utilizing the second "P" narrative, the Prospect Story. Then, ask questions about their perspective around those goals. This approach not only enables you to explore their areas of interest but also provides them with a range of options to choose from, rather than relying solely on "what are you struggling with" questions that could lead you down unexpected paths.

Action Steps:

- For each customer segment you serve, document the top three to five goals/objectives they typically have.
- Create a "day in the life" visualization for your typical customers.
- Prepare open-ended questions that help them elaborate on these goals.
- Use AI research tools to gather industry-specific insights that demonstrate your understanding.

#3. Once you've established the prospect's goals, you have to effectively introduce narrative #3, the Problem Story.

This will be the villain that prevents your hero (prospect) from accomplishing their mission (goals). This Problem Story should be what puts their current status quo at risk. It should speak directly to what could prevent them from accomplishing the goals discussed in the prospect narrative. In addition, you should ensure you "quantify" the problem(s). This gives you the anchor to the cost of the problem that you will want the prospect to compare to later versus the price of your solution. Failure to quantify the problem is a common but costly mistake. Don't make it.

Action Steps:

- Create a catalog of common problems your customers face, organized by industry and role.
- Develop metrics and formulas to quickly quantify these problems in financial terms.
- Build a library of insightful statistics and third-party research that validates these problems.
- Practice asking questions that help the customer self-discover the impact of these problems.

#4. Now that you've connected with your personal "Why" Story, gained alignment on their goals in the Prospect Story, positioned the Problem Story as the barrier to accomplishing those goals, and quantified the cost of either the problem or status quo, it's time for the Product/Solution Story.

Finally, it's your time to solve the problem with your solution. The difference now, however, is that your solution actually looks like a solution since it's solving an actual problem compared to simply features and benefits of a product they really don't care about. Make sure when you present your Product/Solution Story that you explain specifically and simply how it solves the problem you've been discussing.

Action Steps:
- Map your solution capabilities directly to the problems you've identified.
- Create visual frameworks that illustrate how your solution addresses specific challenges.
- Develop clear, concise descriptions of your solution that avoid technical jargon.
- Practice transitioning smoothly from problem quantification to solution introduction.

#5. Next, you simply need to allow their neocortex to validate and justify the positive way they feel about your solution with your Proof Story.

This story contains the evidence that what you are claiming is true and how other customers have solved the same or similar problems with your solution as well.

Action Steps:
- Collect and organize customer success stories by industry, company size, and problem type.

- Format these stories using the narrative structure outlined in Chapter Seven.
- Quantify the results achieved whenever possible.
- Practice delivering these stories naturally without sounding rehearsed.

These customer testimonials or "validation" stories ensure the prospect that they aren't alone. It reduces the cortisol and minimizes the potential perception of risk associated with change.

#6. When it comes to change, keep in mind that the brain has to work through several barriers, even when it feels good about potentially moving forward.

Evaluate your prospect's change barriers through the emotional lens of risk divided by value perception. Ensure you empathize with the concerns they may have from a "risk" standpoint. Address as many of those concerns during your actual Product/Solution Story, if possible, as that may help alleviate the concerns around those barriers before they ever become barriers. Empathetically helping a prospect work through the emotion associated with the perception of risk to them and their organization will go a long way in cementing trust and getting to "yes."

Action Steps:
- Catalog common resistance points you encounter and develop empathetic responses.
- Use the Risk > Value equation to diagnose where resistance is coming from.
- Practice "reset, reduce, reestablish" language patterns for different objection types.
- Create a mental checklist to ensure you've addressed all six common barriers to change.

#7. Don't close. Create an environment of trust with the right amount of urgency for your customer to desire your help in solving the problem(s) you can solve. This will create a bias toward action and commitment.

We know that people choose to buy. Nobody ever wants to be sold. Allow your customer to "choose" your solution as the logical choice to solve their problem. Create a partnership agreement and then hit them with that million-dollar, cognitive-bias-driven commitment question, "What would you like to do?" You've earned the right to ask for a commitment, and they feel they are in control of making that commitment. You both win. More often.

Action Steps:

Practice the language of partnership versus pressure.

- Rehearse the "What would you like to do?" moment until it feels natural.
- Prepare for various responses to this question.
- Develop follow-up strategies that reinforce their decision and minimize post-purchase anxiety.

Implementation Timeline

Adopting NeuroSelling isn't an overnight process. Based on our experience with thousands of sales professionals, here's a realistic timeline for implementation:

Weeks 1-2: Foundation Building

- Study and internalize the scientific principles.
- Draft your personal "Why" Story.
- Begin mapping customer goals and problems.
- Observe your current questioning patterns.

Weeks 3-4: Narrative Development

- Refine your five core narratives.
- Practice them until they become natural.
- Record yourself and get feedback.
- Start using elements in low-stakes conversations.

Weeks 5-8: Controlled Application

- Begin using the full methodology in select customer conversations.
- Debrief after each attempt to identify what worked and what didn't.
- Refine your approach based on real-world feedback.
- Continue practicing the areas that feel least comfortable.

Months 3-6: Mastery Development

- Incorporate NeuroSelling principles into all customer interactions.
- Begin mentoring others in the methodology.
- Track results and identify patterns in successful conversations.
- Continuously refine your narratives based on customer responses.

Common Implementation Challenges and Solutions

Challenge #1: "My personal story doesn't seem relevant to business."

Solution: Remember that your story isn't about the specific details but about the universal beliefs and values it represents. Focus on beliefs that connect to how you serve customers today. Everyone has a story that shaped who they are professionally.

Challenge #2: "I feel awkward asking about emotions in a B2B setting."

Solution: You don't need to directly ask about feelings. Instead, ask about priorities, concerns, and what success looks like. The emotional content will emerge naturally in their responses when there's trust.

Challenge #3: "I'm struggling to quantify the problem effectively."

Solution: Start by identifying one measurable impact of the problem—time, money, opportunity cost, etc. Then, ask the customer to help you understand how that translates to their business. They'll often quantify it for you or give you the variables you need.

Challenge #4: "My company requires me to use standard presentations."

Solution: You can still use the required materials while adapting your narrative approach. Begin with your Why Story before launching the presentation, use the slides as visual support rather than a script, and weave in questions that connect to your customer's specific situation.

Challenge #5: "I'm not seeing immediate results."

Solution: NeuroSelling is a significant shift for most salespeople and requires practice to become natural. Track small wins along the way—longer meetings, more engaged conversations, fewer objections. The big results will follow as you master the approach.

Measuring Your NeuroSelling Effectiveness

How do you know if you're successfully implementing the methodology? Look for these indicators:

1. Conversation Metrics

- Increased talk ratio from customers (they're sharing more)
- More emotionally revealing language from prospects
- Fewer "we need to think about it" responses
- Shorter time to commitment

2. Relationship Indicators

- Prospects share personal information more readily

- Customers refer you to colleagues unprompted
- You're invited into strategic conversations beyond your initial contact
- Prospects reach out between formal meetings with questions or insights

3. Business Results
- Shorter sales cycles
- Higher conversion rates
- Reduced price sensitivity
- Larger initial deals
- Increased expansion in existing accounts

NeuroSelling in Different Contexts

The principles of NeuroSelling apply across a wide range of business contexts, but the specific implementation may vary:

Enterprise B2B Sales
- Focus on building trust across multiple stakeholders.
- Develop role-specific versions of your five narratives.
- Address organizational change resistance more prominently.
- Use Proof Stories that highlight similar complex implementations.

Small Business Sales
- Emphasize personal trust even more strongly.
- Focus on immediate, practical problem solutions.
- Address financial risk barriers directly.
- Use Proof Stories from businesses of similar size and constraints.

Inside Sales/Remote Selling

- Compensate for lack of physical presence with stronger narrative visualization.
- Use technology thoughtfully to enhance connection.
- Be more explicit about summarizing and checking understanding.
- Create more frequent trust-building touchpoints.

Account Management/Customer Success

- Adapt your Why Story to emphasize long-term partnership.
- Focus Problem Stories on unrealized value and new challenges.
- Use Proof Stories that highlight growth trajectories with similar customers.
- Address change resistance around expanding current implementations.

Your First 30 Days with NeuroSelling

If you're ready to get started, here's a thirty-day implementation plan that has proven effective for thousands of our clients:

Days 1-5: Learning and Preparation
- Review the key concepts in this book.
- Draft your personal Why Story.
- Identify the top three customer segments you serve.
- Research their typical goals and challenges.

Days 6-10: Narrative Development
- Refine your Why Story and practice delivering it.
- Develop Prospect Stories for your top customer segments.
- Create Problem Stories that highlight key challenges.
- Catalog your most compelling Proof Stories.

Days 11-15: Practice and Feedback
- Role-play your narratives with colleagues.
- Record yourself delivering key components.
- Get feedback and refine your approach.
- Identify your personal areas for improvement.

Days 16-20: Controlled Implementation
- Select two to three upcoming customer conversations for initial implementation.
- Prepare thoroughly for these conversations.
- Implement the methodology, focusing on personal connection and understanding.
- Immediately debrief what worked and what didn't.

Days 21-30: Refinement and Expansion
- Adjust your approach based on initial feedback.
- Expand to more customer conversations.
- Continue practicing areas that feel uncomfortable.
- Track results and celebrate small wins.

Keep in mind that *sales isn't about selling*.

It's about serving your customers by solving their problems.

That mindset shift will go a long, long way to allowing your prospects and customers to actually believe you care about them. Because you do. In fact, I'll be so bold as to suggest it's even "why" you do what you do. And when you live from your place of "why," your "what" becomes more relevant, and your "how" becomes more easily executed, and your purpose becomes even more evident. Now, go change the world, one problem at a time!

For more information on NeuroSelling, go to
www.braintrustgrowth.com/neuroselling

ENDNOTES

CHAPTER 1

1 Lori Wizdo, "The Ways And Means Of B2B Buyer Journey Maps," Forrester, https://www.forrester.com/blogs/the-ways-and-means-of-b2b-buyer-journey-maps-were-going-deep-at-forresters-b2b-forum.

2 McKinsey & Company, "B2B Decision Maker Pulse Survey," 2023, https://www.mckinsey.com/capabilities/growth-marketing-and-sales/our-insights/these-eight-charts-show-how-covid-19-has-changed-b2b-sales-forever.

3 LinkedIn, "State of Sales Report," 2022, https://business.linkedin.com/sales-solutions/the-state-of-sales-2022-report.

4 Antonio Damasio, "Descartes' Error: Emotion, Reason, and the Human Brain," 1994, https://doi.org/10.1136/bmj.310.6988.1213.

5 Kahneman, Daniel, *Thinking, Fast and Slow* (New York: Farrar, Straus and Giroux, 2011).

6 "The Reality of Information's Impact on B2B Sales," Gartner, https://www.gartner.com/smarterwithgartner/the-reality-informations-impact-b2b-sales, accessed March 30, 2025/.

7 HubSpot Research, "How to Build Trust Online," https://blog.hubspot.com/marketing/more-trustworthy-website.

8 Gartner, "The B2B Buying Journey," 2022, https://www.gartner.com/en/sales/insights/b2b-buying-journey.

9 Barry Schwartz, "The Paradox of Choice: Why More Is Less," Ecco, 2004, https://www.harpercollins.com/products/the-paradox-of-choice-barry-schwartz.

10 Joseph LeDoux, *The Emotional Brain: The Mysterious Underpinnings of Emotional Life* (New York: Simon & Schuster, 1996).

11 David Eagleman, *Incognito: The Secret Lives of the Brain* (New York: Pantheon, 2011).

12 Hilke Plassmann et al., "Branding the Brain: A Critical Review and Outlook," *Journal of Consumer Psychology* 22, no. 1 (2012): 18–36, https://doi.org/10.1016/j.jcps.2011.11.010.

13 Damasio, Antonio, "The Somatic Marker Hypothesis and the Possible Functions of the Prefrontal Cortex," *Philosophical Transactions of the Royal Society*, 1996, https://doi.org/10.1098/rstb.1996.0125.

14 Adamson, Brent and Dixon, Matthew, "The New Sales Imperative," *Harvard Business Review*, 2017, https://hbr.org/2017/03/the-new-sales-imperative.

15 Forrester, "The State of B2B Sales," 2023, https://www.forrester.com/report/the-state-of-b2b-sales-in-2023/RES178687.

16 Gartner, "New B2B Buying Journey & Its Implication for Sales," 2023, https://www.gartner.com/en/sales/insights/b2b-buying-journey.

17 Lisa Feldman Barrett, *How Emotions Are Made: The Secret Life of the Brain* (New York: Houghton Mifflin Harcourt, 2017).

18 Chun Siong Soon et al., "Unconscious Determinants of Free Decisions in the Human Brain," *Nature Neuroscience* 11 (2008): 543–545, https://doi.org/10.1038/nn.2112.

CHAPTER 2

1 Uri Hasson, et al., "Brain-to-Brain Coupling: A Mechanism For Creating And Sharing A Social World," *Trends in Cognitive Sciences* 16, no. 2 (2012): 114–121, https://doi.org/10.1016/j. tics.2011.12.007.

2 Paul J. Zak, "Why Inspiring Stories Make Us React: The Neuroscience of Narrative," *Cerebrum* 2, (2015), https://www.ncbi.nlm.nih.gov/pmc/articles/PMC4445577/.

3 Antonio R. Damasio, "The Somatic Marker Hypothesis and the Possible Functions of the Prefrontal Cortex," *Philosophical Transactions of the Royal Society of London* 351, no. 1346 (1996), https://doi.org/10.1098/rstb.1996.0125

4 Carmen Simon, *Impossible to Ignore: Creating Memorable Content to Influence Decisions* (New York: McGraw-Hill Education, 2016).

5 Mara Mather, "Emotional Arousal and Memory Binding: An Object-Based Framework," *Perspectives on Psychological Science* 2, no. 1 (2007): 33–52, https://pubmed.ncbi.nlm.nih. gov/26151918/

6 "Baba Shiv: Emotions Can Negatively Impact Investment Decisions," Stanford Business, September 1, 2005, https://www.gsb.stanford.edu/insights/ baba-shiv-emotions-can-negatively-impact-investment-decisions.

7 Elizabeth A. Kensinger and Suzanne Corkin, "Memory Enhancement for Emotional Words: Are Emotional Words More Vividly Remembered Than Neutral Words?" *Memory & Cognition* 31, no. 8 (2003): 1169–1180, https://doi.org/10.3758/BF03195800.

8 Dan Ariely, *Predictably Irrational: The Hidden Forces That Shape Our Decisions* (New York: Harper Perennial, 2010).

9 Robert B. Cialdini, *Pre-Suasion: A Revolutionary Way to Influence and Persuade* (New York: Simon & Schuster, 2016).

10 Amy J. Cuddy, Susan T. Fiske, and Peter Glick, "Warmth and Competence as Universal Dimensions of Social Perception: The Stereotype Content Model and the BIAS Map," *Advances in Experimental Social Psychology* 40 (2008): 61–149, https://doi.org/10.1016/ S0065-2601(07)00002-0.

11 Susan Fournier and Claudio Alvarez, "White Paper: Brands as Relationships Partners," Fidelum Partners, https://fidelum.com/whitepapers/brands-relationship-partners.

CHAPTER 3

1 Paul D. MacLean, *The Triune Brain in Evolution: Role in Paleocerebral Functions* (New York: Plenum, 1990).

2 Daniel Kahneman, *Thinking, Fast and Slow* (New York: Macmillan, 2011).

3 Mara Mather and Marcia K. Johnson, "Choice-Supportive Source Monitoring: Do Our Decisions Seem Better to Us as We Age?" Psychology and Aging 15, no. 4 (2000): 596–606, https://doi.org/10.1037/0882-7974.15.4.596.

4 Hugo Mercier and Dan Sperber, "Why Do Humans Reason? Arguments For an Argumentative Theory," *Behavioral and Brain Sciences*, November 2020, https://doi.org/10.1017/ S0140525X10000968.

5   Michael D. Fox et al., "The Human Brain Is Intrinsically Organized into Dynamic, Anticorrelated Functional Networks," *Proceedings of the National Academy of Sciences* 102, no. 27 (2005): 9673–9678, https://doi.org/10.1073/pnas.0504136102.

6   Richard E. Boyatzis, Kylie Rochford, and Anthony I. Jack, "Antagonistic Neural Networks Underlying Differentiated Leadership Roles," *Frontiers in Human Neuroscience* 8, 2014, https://doi.org/10.3389/fnhum.2014.00114.

7   Anthony I. Jack et al., "fMRI Reveals Reciprocal Inhibition Between Social and Physical Cognitive Domains," *NeuroImage* 66, no. 1 (2013): 385–401, https://doi.org/10.1016/j.neuroimage.2012.10.061.

8   Ralph Adolphs, "The Social Brain: Neural Basis Of Social Knowledge," *Annual Review of Psychology* 60 (2009): 693–716, https://doi.org/10.1146/annurev.psych.60.110707.163514.

9   Paul J. Whalen et al., "Masked Presentations of Emotional Facial Expressions Modulate Amygdala Activity Without Explicit Knowledge," *Journal of Neuroscience* 18, no. 1 (1998): 411–418, https://doi.org/10.1523/JNEUROSCI.18-01-00411.1998.

10 Amy F. T. Arnsten, "Stress Signalling Pathways That Impair Prefrontal Cortex Structure and Function," *Nature Reviews Neuroscience* 10 (2009):410-422, https://doi.org/10.1038/nrn2648.

11 Lars Schwabe and Oliver T. Wolf, "Stress and Multiple Memory Systems: From 'Thinking' to 'Doing,'" *Trends in Cognitive Sciences* 17, no. 2 (2013): 60–68, https://pubmed.ncbi.nlm.nih.gov/23290054/.

12 Livia Tomova et al., "Is Stress Affecting Our Ability to Tune into Others? Evidence for Gender Differences in the Effects of Stress on Self-Other Distinction," *Psychoneuroendocrinology* 43 (2014): 95–104, https://pubmed.ncbi.nlm.nih.gov/24703175/.

13 Anthony J. Porcelli and Mauricio R. Delgado, "Acute Stress Modulates Risk-Taking in Financial Decision-Making," *Psychological Science* 20, no. 3 (2009): 278–283, https://pmc.ncbi.nlm.nih.gov/articles/PMC4882097/.

14 Peter Sokol-Hessner et al., "Emotion Regulation Reduces Loss Aversion and Decreases Amygdala Responses to Losses," *Social Cognitive and Affective Neuroscience* 8, no.3 (2013): 341–350, https://doi.org/10.1093/scan/nss002.

CHAPTER 4

1   Edelman, "2023 Edelman Trust Barometer," https://www.edelman.com/trust/2023/trust-barometer.

2   Edelman, "Trust Barometer."

3   "State of the Connected Customer," Salesforce, 2022, https://www.salesforce.com/resources/research-reports/state-of-the-connected-customer/.

4   "Future of Customer Experience," PwC, 2022, https://www.pwc.com/us/en/services/consulting/library/consumer-intelligence-series/future-of-customer-experience.html.

5   Lisa Donchak et al, "The Future of B2B Sales is Hybrid," McKinsey & Company, April 27, 2022, https://www.mckinsey.com/capabilities/growth-marketing-and-sales/our-insights/the-future-of-b2b-sales-is-hybrid. .

6   Michael Kosfeld et al., "Oxytocin Increases Trust in Humans," *Nature* 435 (2005): 673–676, https://www.nature.com/articles/nature03701.

7   Paul J. Zak, "The Neuroscience of Trust," *Harvard Business Review* 95, no. 1 (2017): 84–90, https://hbr.org/2017/01/the-neuroscience-of-trust.

8   Thomas Baumgartner et al., "Oxytocin Shapes the Neural Circuitry of Trust and Trust Adaptation In Humans," *Neuron* 58, no. 4 (2008): 639–650, https://www.cell.com/neuron/fulltext/S0896-6273(08)00327-9.

9   Brené Brown, *Daring Greatly: How the Courage to Be Vulnerable Transforms the Way We Live, Love, Parent, and Lead* (New York: Gotham Books, 2012), 34.

10 Antonio Damasio and Gil. B. Carvalho, "The Nature Of Feelings: Evolutionary And Neurobiological Origins," *Nature Reviews Neuroscience* 14 (2013):143–152, https://www.nature.com/articles/nrn3403.

11 Laura Delizonna, "High-Performing Teams Need Psychological Safety: Here's How to Create It," Harvard Business Review, August 24, 2017, https://hbr.org/2017/08/high-performing-teams-need-psychological-safety-heres-how-to-create-it..

12 Alexander Todorov, Manish Pakrashi, and Nikolaas N. Oosterhof, "Evaluating Faces on Trustworthiness After Minimal Time Exposure," *Social Cognition* 27, no. 6 (2009): 813–833, https://doi.org/10.1521/soco.2009.27.6.813.

13 Peter H. Kim, Donald L. Ferrin, Cecily D. Cooper, and Kurt T. Dirks, "Removing the Shadow Of Suspicion: The Effects of Apology Versus Denial For Repairing Competence—Versus Integrity-Based Trust Violations," *Journal of Applied Psychology* 89, no. 1 (2004): 104–118, https://doi.org/10.1037/0021-9010.89.1.104.

14 Uri Hasson et al., "Brain-to-Brain Coupling: A Mechanism for Creating and Sharing a Social World," *Trends in Cognitive Sciences* 16, no. 2 (2012): 114–121, https://www.cell.com/trends/cognitive-sciences/fulltext/S1364-6613(11)00242-8.

15 Melanie C. Green and Timothy C. Brock, "The Role of Transportation in the Persuasiveness of Public Narratives," *Journal of Personality and Social Psychology* 79, no. 5 (2000): 701–721, https://psycnet.apa.org/record/2000-05929-004.

16 Green, "The Role of Transportation."

17 Amy C. Edmondson and Zhike Lei, "Psychological Safety: The History, Renaissance, and Future of an Interpersonal Construct," *Annual Review of Organizational Psychology and Organizational Behavior* 1, no. 1 (2014): 23–43, https://www.annualreviews.org/doi/10.1146/annurev-orgpsych-031413-091305.

18 Robert Goffee and Gareth Jones, "Why Should Anyone Be Led by You?" *Harvard Business Review* 78, no. 5, (2000): 62–70. https://hbr.org/2000/09/why-should-anyone-be-led-by-you.

19 Amotz Zahavi and Avishag Zahavi, The Handicap Principle: A Missing Piece of Darwin's Puzzle (Oxford: Oxford University Press, 1997).

20 Brené Brown, *Dare to Lead: Brave Work. Tough Conversations. Whole Hearts* (New York: Random House, 2018).

21 Joseph Henrich, *The Secret of Our Success: How Culture Is Driving Human Evolution, Domesticating Our Species, and Making Us Smarter* (New Jersey: Princeton University Press, 2015).

22 Mary Helen Immordino-Yang, Andrea McColl, Hanna Damasio, and Antonio Damasio, "Neural Correlates of Admiration and Compassion," *Proceedings of the National Academy of Sciences* 106, no. 19 (2009): 8021–8026, https://www.pnas.org/content/106/19/8021.

23 Matthew D. Lieberman, *Social: Why Our Brains Are Wired to Connect* (New York: Crown, 2013).

24 Ruth Feldman, Aron Weller, and Ari Levine, "Evidence for a Neuroendocrinological Foundation of Human Affiliation: Plasma Oxytocin Levels Across Pregnancy and the Postpartum Period Predict Mother-Infant Bonding," *Psychological Science* 18, no. 11 (2007): 965–970, https://journals.sagepub.com/doi/10.1111/j.1467-9280.2007.02010.x.

25 Ilanit Gordon et al., "Oxytocin and the Development of Parenting in Humans," *Biological Psychiatry* 68, no. 4 (2010): 377–382, https://pmc.ncbi.nlm.nih.gov/articles/PMC3943240/.

26 Miho Nagasawa et al., "Dog's Gaze at Its Owner Increases Owner's Urinary Oxytocin During Social Interaction," *Hormones and Behavior* 55, no. 3 (2009): 434–441, https://pubmed.ncbi.nlm.nih.gov/19124024/.

27 Peter Kirsch et al., "Oxytocin Modulates Neural Circuitry for Social Cognition and Fear in Humans," *Journal of Neuroscience* 25, no. 49 (2005): 11489–11493. https://www.jneurosci.org/content/25/49/11489.

28 Markus Heinrichs et al., "Social Support and Oxytocin Interact to Suppress Cortisol and Subjective Responses to Psychosocial Stress," *Biological Psychiatry* 54, no. 12 (2003): 1389–1398, https://www.biologicalpsychiatryjournal.com/article/S0006-3223(03)00465-7/fulltext.

29 Angeliki Theodoridou et al., "Oxytocin and Social Perception: Oxytocin Increases Perceived Facial Trustworthiness and Attractiveness," *Hormones and Behavior* 56, no. 1 (2009): 128–132, https://www.sciencedirect.com/science/article/abs/pii/S0018506X09000853.

30 Thomas Baumgartner et al., "Oxytocin Shapes the Neural Circuitry of Trust and Trust Adaptation in Humans," *Neuron* 58, no. 4 (2008): 639–650, https://www.cell.com/neuron/fulltext/S0896-6273(08)00327-9.

31 Paul J. Zak, *Trust Factor: The Science of Creating High-Performance Companies* (New York: AMACOM, 2017).

32 Paul J. Zak and Stephen Knack, "Trust and Growth," *The Economic Journal* 111, no. 470 (2001): 295–321, https://onlinelibrary.wiley.com/doi/abs/10.1111/1468-0297.00609.

33 Jorge A. Barraza and Paul J. Zak, "Empathy Toward Strangers Triggers Oxytocin Release and Subsequent Generosity," *Annals of the New York Academy of Sciences* 1167, no. 1 (2009): 182–189, https://nyaspubs.onlinelibrary.wiley.com/doi/abs/10.1111/j.1749-6632.2009.04504.x.

34 Paul J. Zak, *The Moral Molecule: How Trust Works* (New York: Plume, 2013).

35 Amy J. C. Cuddy et al., "Connect, Then Lead," Harvard Business Review 91, no. 7 (2013): 54–61, https://hbr.org/2013/07/connect-then-lead.

36 Chad Boutin, "Snap Judgments Decide a Face's Character, Psychologist Finds," *Princeton University News*, August 22, 2006, https://www.princeton.edu/news/2006/08/22/snap-judgments-decide-faces-character-psychologist-finds.

37 Diana I. Tamir and Jason P. Mitchell, "Disclosing Information About the Self Is Intrinsically Rewarding," *Proceedings of the National Academy of Sciences* 109, no. 21 (2012): 8038–8043, https://www.researchgate.net/publication/224919007_Disclosing_information_about_the_self_is_intrinsically_rewarding.

CHAPTER 5

1 David Rock and Jeffrey Schwartz, "The Neuroscience of Leadership," *Strategy+Business* 43, 71–79, https://www.strategy-business.com/article/06207.

2 Daniel Kahneman, Jack L. Knetsch, and Richard H. Thaler, "Anomalies: The Endowment Effect, Loss Aversion, and Status Quo Bias," *Journal of Economic Perspectives* 5, no. 1 (1991): 193–206, https://doi.org/10.1257/jep.5.1.193.

3 William Samuelson and Richard Zeckhauser, "Status Quo Bias in Decision-Making," *Journal of Risk and Uncertainty* 1, no. 1 (1988): 7—59, https://doi.org/10.1007/BF00055564.

4 Matthew D. Lieberman and Naomi I. Eisenberger, "Pains and Pleasures of Social Life," *Science* 323, no. 5916 (2009): 890–891, https://doi.org/10.1126/science.1170008.

5 Kurt Lewin, "Frontiers in Group Dynamics: Concept, Method and Reality in Social Science; Social Equilibria and Social Change," *Human Relations* 1, no. 1 (1947): 5–41, https://doi.org/10.1177/001872674700100103.

6 Gary Keller and Jay Papasan, *The ONE Thing: The Surprisingly Simple Truth Behind Extraordinary Results* (New York, Bard Press, 2013), 173.

7 Amy F. T. Arnsten, "Stress Signaling Pathways That Impair Prefrontal Cortex Structure and Function," *Nature Reviews Neuroscience* 10, no. 6 (2009): 410–422, https://doi.org/10.1038/nrn2648.

8 Naomi I. Eisenberger et al., "Does Rejection Hurt? An fMRI Study of Social Exclusion," *Science* 302, no. 5643 (2003): 290–292, https://doi.org/10.1126/science.1089134.

9 Julianne Holt-Lunstad et al., "Social Relationships and Mortality Risk: A Meta-Analytic Review," *PLoS Medicine* 7, no. 7 (2010): e1000316, https://doi.org/10.1371/journal.pmed.1000316.

10 Deborah A. Prentice and Dale T. Miller, "Pluralistic Ignorance and Alcohol Use on Campus: Some Consequences of Misperceiving the Social Norm," *Journal of Personality and Social Psychology* 64, no. 2 (1993): 243–256, https://doi.org/10.1037/0022-3514.64.2.243.

11 Amos Tversky and Daniel Kahneman, "Advances in Prospect Theory: Cumulative Representation of Uncertainty," *Journal of Risk and Uncertainty* 5, no. 4 (1992): 297–323, https://doi.org/10.1007/BF00122574.

12 Richard Thaler, "Toward a Positive Theory of Consumer Choice," *Journal of Economic Behavior & Organization* 1, no. 1 (1980): 39–60, https://doi.org/10.1016/0167-2681(80)90051-7.

13 Barry Schwartz, *The Paradox of Choice: Why More Is Less* (New York: Harper Perennial, 2016).

14 Russell A. Poldrack and Mark G. Packard, "Competition Among Multiple Memory Systems: Converging Evidence From Animal And Human Brain Studies," *Neuropsychologia* 41, no. 3 (2003): 245–251, https://doi.org/10.1016/S0028-3932(02)00157-4.

15 Sendhil Mullainathan and Eldar Shafir, *Scarcity: Why Having Too Little Means So Much* (New York: Macmillan, 2014).

16 Anandi Mani et al., "Poverty impedes cognitive function," *Science* 341, no. 6149 (2013): 976–980, https://doi.org/10.1126/science.1238041.

17 Jay Schulkin, "Allostasis, Homeostasis, and the Costs of Physiological Adaptation," *Cambridge University Press*, 2004, https://doi.org/10.1017/CBO9781316257081.

18 Practical Psychology, reviewed by Kristen Clure, MA, "Functional Fixedness," https://practicalpie.com/functional-fixedness/.

19 Thaler, Richard H. Misbehaving: The Making of Behavioral Economics. New York: W.W. Norton & Company, 2016.

20 Elisabeth Kübler-Ross and David Kessler, On Grief and Grieving: Finding the Meaning of Grief Through the Five Stages of Loss (New York: Simon and Schuster, 2014).

21 William Bridges, PhD, Managing Transitions: Making the Most of Change (New York: Da Capo Press, 1991).

CHAPTER 6

[1] Martie G. Haselton and Daniel Nettle, "The Paranoid Optimist: An Integrative Evolutionary Model of Cognitive Biases," *Personality and Social Psychology Review* 10, no. 1 (2006): 47–66, https://doi.org/10.1207/s15327957pspr1001_3.

2 Gerd Gigerenzer et al., *Simple Heuristics That Make Us Smart* (Oxford: Oxford University Press, 1999).

3 Sabrina M. Tom et al., "The Neural Basis of Loss Aversion in Decision-Making Under Risk," *Science* 315, no. 5811 (2007): 515–518, https://doi.org/10.1126/science.1134239.

4 Paul J. Zak, *Trust Factor: The Science of Creating High-Performance Companies* (New York: AMACOM, 2017).

5 Daniel Kahneman et al., "Before You Make That Big Decision," *Harvard Business Review* 89, no. 6 (2011): 50–60, https://hbr.org/2011/06/the-big-idea-before-you-make-that-big-decision.

6 Adrian Furnham and Hua Chu Boo, "A Literature Review of the Anchoring Effect," *The Journal of Socio-Economics* 40, no. 1 (2011): 35–42, https://doi.org/10.1016/j.socec.2010.10.006.

7 Nicholas Epley and Thomas Gilovich, "When Effortful Thinking Influences Judgmental Anchoring: Differential Effects of Forewarning and Incentives on Self-Generated and Externally Provided Anchors," *Journal of Behavioral Decision-Making* 18, no. 3 (2005): 199–212, https://doi.org/10.1002/bdm.495.

8 Lus Wathieu and Marco Bertini, "Price as a Stimulus to Think: The Case for Willful Overpricing," *Marketing Science* 26, no. 1 (2007): 118–129, https://www.jstor.org/stable/40057077.

9 Mara Mather and Marcia K. Johnson, "Choice-Supportive Source Monitoring: Do Our Decisions Seem Better to Us as We Age?" *Psychology and Aging* 15, no. 4 (2000): 596–606, https://doi.org/10.1037/0882-7974.15.4.596.

10 Linda A. Henkel and Mara Mather, "Memory Attributions for Choices: How Beliefs Shape Our Memories," *Journal of Memory and Language* 57, no. 2 (2007): 163–176, https://doi.org/10.1016/j.jml.2006.08.012.

11 Johanna M. Jarcho et al., "The Neural Basis of Rationalization: Cognitive Dissonance Reduction During Decision-Making," *Social Cognitive and Affective Neuroscience* 6, no. 4 (2010): 460–467, https://doi.org/10.1093/scan/nsq054.

12 Elizabeth A. Phelps, "Human Emotion and Memory: Interactions of the Amygdala and Hippocampal Complex," *Current Opinion in Neurobiology* 14, no. 2 (2004): 198–202, https://doi.org/10.1016/j.conb.2004.03.015.

13 James D. Westphal and Edward J. Zajac, "Defections from the Inner Circle: Social Exchange, Reciprocity, and the Diffusion of Board Independence In US Corporations," *Administrative Science Quarterly* 42, no. 1 (1997): 161–183, https://doi.org/10.2307/2393812.

14 Zakary L. Tormala and Joshua J. Clarkson, "Assimilation and Contrast in Persuasion: The Effects of Source Credibility in Multiple Message Situations," *Personality and Social Psychology Bulletin* 33, no. 4 (2007): 559–571, https://doi.org/10.1177/0146167206296955.

15 Drew Westen et al., "Neural Bases of Motivated Reasoning: An fMRI Study of Emotional Constraints on Partisan Political Judgment in the 2004 US Presidential Election," *Journal of Cognitive Neuroscience* 18, no. 11 (2006): 1947–1958, https://doi.org/10.1162/jocn.2006.18.11.1947.

16 Joseph T. Klapper, The Effects of Mass Communication (Los Angeles: Free Press, 1960).

17 Brendan Nyhan and Jason Reifler, "When Corrections Fail: The Persistence of Political Misperceptions," *Political Behavior* 32, no. 2(2010): 303–330, https://doi.org/10.1007/s11109-010-9112-2.

18 Jonas T. Kaplan et al., "Neural Correlates of Maintaining One's Political Beliefs in the Face of Counterevidence," *Scientific Reports* 6, 39589 (2016), https://doi.org/10.1038/srep39589.

19 Amos Tversky and Daniel Kahneman, "Availability: A Heuristic for Judging Frequency and Probability," *Cognitive Psychology* 5, no. 2 (1973): 207–232, https://doi.org/10.1016/0010-0285(73)90033-9.

20 Robert J. Shiller, *Irrational Exuberance* (New Jersey: Princeton University Press, 2000).

21 Chris D. Frith and Uta Frith, "Implicit and Explicit Processes in Social Cognition," *Neuron* 60, no. 3 (2008): 503–510, https://doi.org/10.1016/j.neuron.2008.10.032.

22 Vasily Klucharev et al., "Reinforcement Learning Signal Predicts Social Conformity," *Neuron* 61, no. 1 (2009): 140–151, https://doi.org/10.1016/j.neuron.2008.11.027.

23 Gregory S. Berns et al., "Neurobiological Correlates of Social Conformity and Independence During Mental Rotation," *Biological Psychiatry* 58, no. 3 (2005): 245–253, https://doi.org/10.1016/j.biopsych.2005.04.012.

24 Adam L. Alter and Daniel L. Oppenheimer, "Uniting the Tribes of Fluency to Form a Metacognitive Nation," *Personality and Social Psychology Review* 13, no. 3 (2009): 219–235, https://doi.org/10.1177/1088868309341564.

25 "B2B Buying: How Top CSOs and CMOs Optimize the Journey," Gartner Research, https://www.gartner.com/en/sales/insights/b2b-buying-journey.

CHAPTER 7

1 Evelina Fedorenko and Sharon L. Thompson-Schill, "Reworking the Language Network," *Trends in Cognitive Sciences* 18, no. 3 (2014): 120–126, https://doi.org/10.1016/j.tics.2013.12.006.

2 Uri Hasson, Janice Chen, and Christopher J. Honey, "Hierarchical Process Memory: Memory as an Integral Component of Information Processing," *Trends in Cognitive Sciences* 19, no. 6 (2015): 304–313, https://doi.org/10.1016/j.tics.2015.04.006.

3 World Bank, "World Development Report 2015: Mind, Society, and Behavior," Washington, DC: World Bank, https://www.worldbank.org/en/publication/wdr2015.

4   Jennifer Aaker, "Harnessing the Power of Stories," VMware Women's Leadership Innovation Lab, Stanford University, accessed April 9, 2025, https://womensleadership.stanford.edu/node/796/harnessing-power-stories.

5   Paul J. Zak, "Why Your Brain Loves Good Storytelling," *Harvard Business Review*, October 28, 2014, https://hbr.org/2014/10/why-your-brain-loves-good-storytelling.

6   John Medina, *Brain Rules: Twelve Principles for Surviving and Thriving at Work, Home, and School Pear Press* (Washington: Pear Press, 2008).

7   Greg J. Stephens, Lauren J. Silbert, and Uri Hasson, "Speaker-Listener Neural Coupling Underlies Successful Communication," *Proceedings of the National Academy of Sciences* 107, no. 32 (2010): 14425–14430, https://doi.org/10.1073/pnas.1008662107.

8   Uri Hasson et al., "Brain-to-Brain Coupling: A Mechanism for Creating and Sharing a Social World," *Trends in Cognitive Sciences* 16, no. 2 (2012): 114–121, https://doi.org/10.1016/j.tics.2011.12.007.

9   Anna Lundqvist et al., "The Impact of Storytelling on the Consumer Brand Experience: The Case of a Firm-Originated Story," *Journal of Brand Management* 20, no. 4 (2013): 283–297, https://doi.org/10.1057/bm.2012.15.

10 George A. Miller, "The Magical Number Seven, Plus or Minus Two: Some Limits on Our Capacity For Processing Information," *Psychological Review* 63, no. 2 (1956): 81–97, https://doi.org/10.1037/h0043158.

11 Daniel J. Levitin, *The Organized Mind: Thinking Straight in the Age of Information Overload* (New York: Dutton, 2014).

12 Michael L. Anderson, "Neural Reuse: A Fundamental Organizational Principle of the Brain," *Behavioral and Brain Sciences* 33, no. 4 (2010): 245–266, https://doi.org/10.1017/S0140525X10000853.

13 Simon Lacey, Randall Stilla, K. Sathian, "Metaphorically Feeling: Comprehending Textural Metaphors Activates Somatosensory Cortex," *Brain and Language* 120, no. 3 (2012): 416–421, https://doi.org/10.1016/j.bandl.2011.12.016.

14 Allan Paivio, *Mental Representations: A Dual Coding Approach* (Oxford: Oxford University Press, 1990).

15 Daniel L. Schacter and Donna Rose Addis, "The Cognitive Neuroscience of Constructive Memory: Remembering the Past and Imagining the Future," *Philosophical Transactions of the Royal Society* B: Biological Sciences 362, no. 1481 (2007): 773–786, https://doi.org/10.1098/rstb.2007.2087.

16 Jerome Bruner, *Actual Minds, Possible Worlds* (Boston: Harvard University Press, 1986).

17 Joseph Campbell, The Hero with a Thousand Faces, Third Edition (California: New World Library, 2008).

18 Sharon S. Brehm and Jack W. Brehm, *Psychological Reactance: A Theory of Freedom and Control* (Cambridge, MA: Academic Press, 2013).

19 Nicole K. Speer et al., "Reading Stories Activates Neural Representations of Visual and Motor Experiences," *Psychological Science* 20, no. 8 (2009): 989–999, https://doi.org/10.1111/j.1467-9280.2009.02397.x.

CHAPTER 8

1 Chris D. Frith and Uta Frith, "The Neural Basis of Mentalizing," *Neuron* 50, no. 4 (2006): 531–534, https://doi.org/10.1016/j.neuron.2006.05.001.

2 David Mayer and Herbert M. Greenberg, "What Makes a Good Salesman," *Harvard Business Review* 84, no. 7/8 (2006): 164–171, https://hbr.org/2006/07/what-makes-a-good-salesman.

3 Robert M. Sapolsky, *Behave: The Biology of Humans at Our Best and Worst* (New York: Penguin Press, 2017).

4 Joseph E. LeDoux, *Anxious: Using the Brain to Understand and Treat Fear and Anxiety* (New York: Viking, 2015).

5 Aviva Philipp Muller et al., "Understanding When Similarly Induced Affective Attraction Predicts Willingness to Affiliate: An Attitude Strength Perspective," *Frontiers in Psychology* 11 (2020), https://doi.org/10.3389/fpsyg.2020.01919.

6 Radicati Group. "Email Statistics Report, 2015–2019." https://www.radicati.com/wp/wp-content/uploads/2015/02/Email-Statistics-Report-2015-2019-Executive-Summary.pdf.

7 Leslie A. Perlow et al., "Stop the Meeting Madness," *Harvard Business Review*, July–August 2017, https://hbr.org/2017/07/stop-the-meeting-madness.

8 Cara Capuano, "Can't Pay Attention? You're Not Alone," University of California, Irvine, https://www.universityofcalifornia.edu/news/cant-pay-attention-youre-not-alone.

CHAPTER 9

1 Sung-il Kim, "Neuroscientific Model of Motivational Process," *Frontiers in Psychology* 4, no. 98 (2013): https://doi.org/10.3389/fpsyg.2013.00098.

2 Sophie Wohltjen and Thalia Wheatley, "Eye Contact Marks the Rise and Fall of Shared Attention in Conversation," *Proceedings of the National Academy of Sciences of the United States of America* 118, no. 37 (2021): e2106645118, https://www.ncbi.nlm.nih.gov/pmc/articles/PMC8449315.

3 Enhui Xie et al., "Sharing Happy Stories Increases Interpersonal Closeness: Interpersonal Brain Synchronization as a Neural Indicator," *eNeuro* 8, no. 6 (2021), https://www.eneuro.org/content/8/6/ENEURO.0245-21.2021.

4 Dan Wang, Changhong Liu, and Wenfeng Chen, "The Role of Self-Representation in Emotional Contagion," *Frontiers in Human Neuroscience* 18, 2024, https://www.frontiersin.org/articles/10.3389/fnhum.2024.1361368/full.

5 Uri Hasson et al., "Enhanced Intersubject Correlations During Movie Viewing Correlate with Successful Episodic Encoding," *Neuron* 57, no. 3 (2008): 452–462, https://doi.org/10.1016/j.neuron.2007.12.009.

6 Christopher R. Madan, "Exploring Word Memorability: How Well Do Different Word Properties Explain Item Free-Recall Probability?" *Neurobiology of Learning and Memory* 28 (2020): 583–595, https://link.springer.com/article/10.3758/s13423-020-01820-w.

7 Christoph Hofstetter et al., "Reactivation of Visual Cortex During Memory Retrieval: Content Specificity and Emotional Modulation," *NeuroImage* 60, no. 3 (2012): 1734–1745, https://www.sciencedirect.com/science/article/abs/pii/S1053811912001280.

8   Jeremy N. Bailenson, "Nonverbal Overload: A Theoretical Argument for the Causes of Zoom Fatigue," *Technology, Mind, and Behavior* 2, no. 1 (2021), https://doi.org/10.1037/tmb0000030.

9   Nick Morgan, *Can You Hear Me? How to Connect with People in a Virtual World* (Boston, MA: Harvard Business Review Press, 2018).

10 George Northoff et al., "Self-Referential Processing In Our Brain—A Meta-Analysis of Imaging Studies on the Self," *Neuroimage* 31, no. 1 (2006): 440–457, https://doi.org/10.1016/j.neuroimage.2005.12.002.

11 Heather M. Gray et al., "P300 as an Index of Attention to Self-Relevant Stimuli," *Journal of Experimental Social Psychology* 40, no. 2 (2004): 216–224, https://doi.org/10.1016/S0022-1031(03)00092-1.

12 Peter M. Gollwitzer, "Implementation Intentions: Strong Effects of Simple Plans," *American Psychologist* 54, no. 7 (1999): 493–503, https://doi.org/10.1037/0003-066X.54.7.493.

13 Mark J. Landau, Brian P. Meier, and Lucas A. Keefer, "A Metaphor-Enriched Social Cognition," *Psychological Bulletin* 136, no. 6 (2010): 1045–1067, https://doi.org/10.1037/a0020970.

14 Francesca M. M. Citron and Adele E. Goldberg, "Metaphorical Sentences Are More Emotionally Engaging Than Their Literal Counterparts," *Journal of Cognitive Neuroscience* 26, no. 11 (2014): 2585–2595, https://doi.org/10.1162/jocn_a_00654.

15 Maurizio Corbetta and Gordon L. Shulman, "Control of Goal-Directed and Stimulus-Driven Attention in the Brain," *Nature Reviews Neuroscience* 3, no. 3 (2002): 201–215, https://doi.org/10.1038/nrn755.

16 Clay B. Holroyd and Michael G. H. Coles, "The Neural Basis of Human Error Processing: Reinforcement Learning, Dopamine, and the Error-Related Negativity," *Psychological Review* 109, no. 4 (2002): 679–709, https://doi.org/10.1037/0033-295X.109.4.679.

17 Sabrina M. Tom et al., "The Neural Basis of Loss Aversion in Decision-Making Under Risk," *Science* 315, no. 5811 (2007): 515–518, https://doi.org/10.1126/science.1134239.

18 Leon Festinger, *A Theory of Cognitive Dissonance* (California: Stanford University Press, 1957).

19 Robert B. Cialdini, *Influence: The Psychology of Persuasion Revised Edition* (New York: Harper Business, 2006).

20 Amos Tversky and Daniel Kahneman, "Judgment under Uncertainty: Heuristics and Biases," *Science* 185, no. 4157 (1974): 1124–1131, https://doi.org/10.1126/science.185.4157.1124.

21 Wolfram Schultz, "Dopamine Reward Prediction Error Coding," Dialogues in Clinical *Neuroscience* 18, no. 1 (2016): 23–32, https://doi.org/10.31887/DCNS.2016.18.1/wschultz.

22 Daniel L. Schacter, Donna Rose Addis, and Randy L. Buckner, "Remembering the Past to Imagine the Future: The Prospective Brain," *Nature Reviews Neuroscience* 8, no. 9 (2007): 657–661, https://doi.org/10.1038/nrn2213.

23 John Sweller, "Cognitive Load Theory," *Psychology of Learning and Motivation* 55 (2011): 37–76, https://doi.org/10.1016/B978-0-12-387691-1.00002-8.

24 Sheena S. Iyengar and Mark R. Lepper, "When Choice Is Demotivating: Can One Desire Too Much of a Good Thing?" *Journal of Personality and Social Psychology* 79, no. 6 (2000): 995–1006, https://doi.org/10.1037/0022-3514.79.6.995.

25 Cristina M. Atance and Daniela K. O'Neill, "Episodic Future Thinking," *Trends in Cognitive Sciences* 5, no. 12 (2001): 533–539, https://doi.org/10.1016/S1364-6613(00)01804-0.

26 Daniel T. Gilbert and Timothy D. Wilson, "Prospection: Experiencing the Future," *Science* 317, no. 5843 (2007): 1351–1354, https://doi.org/10.1126/science.1144161.

27 Giacomo Rizzolatti and Laila Craighero, "The Mirror-Neuron System," *Annual Review of Neuroscience* 27 (2004): 169–192, https://doi.org/10.1146/annurev.neuro.27.070203.144230.

28 Lindsay M. Oberman et al., "The Human Mirror Neuron System: A Link Between Action Observation and Social Skills," *Social Cognitive and Affective Neuroscience* 2, no. 1 (2007): 62–66, https://doi.org/10.1093/scan/nsl022.

29 Melanie C. Green and Timothy C. Brock, "The Role of Transportation in the Persuasiveness of Public Narratives," *Journal of Personality and Social Psychology* 79, no. 5 (2000): 701–721, https://doi.org/10.1037/0022-3514.79.5.701.

30 Jennifer Aaker, *Harnessing the Power of Stories: The Impact of Narrative on Consumer Behavior* (California: Stanford Graduate School of Business Research Paper, 2018).

31 Hilke Plassmann, John O'Doherty, and Antonio Rangel, "Orbitofrontal Cortex Encodes Willingness To Pay In Everyday Economic Transactions," *Journal of Neuroscience* 27, no. 37 (2007): 9984–9988, https://doi.org/10.1523/JNEUROSCI.2131-07.2007.

32 Brian Knutson et al., "Neural Predictors of Purchases," *Neuron* 53 no. 1 (2007): 147–156, https://doi.org/10.1016/j.neuron.2006.11.010.

CHAPTER 10

1 Judith E. Glaser and Richard D. Glaser, "The Neurochemistry of Positive Conversations," *Harvard Business Review*, June 12, 2014, https://hbr.org/2014/06/the-neurochemistry-of-positive-conversations.

CHAPTER 11

1 Haley V. West et al, "Amygdala Activation in Cognitive Task fMRI Varies with Individual Differences in Cognitive Traits," National Library of Medicine, https://pmc.ncbi.nlm.nih.gov/articles/PMC8480985/.

2 Clay B. Holroyd and Michael G. H. Coles, "The Neural Basis of Human Error Processing: Reinforcement Learning, Dopamine, and the Error-Related Negativity," *Psychological Review* 109, no. 4 (2002): 679–709, https://doi.org/10.1037/0033-295X.109.4.679.

3 Henri Tajfel and John C. Turner, "An Integrative Theory of Intergroup Conflict," in *The Social Psychology of Intergroup Relations*, ed. W.G. Austin & S. Worchel (Stanford University Press, 1979), 33–47.

CHAPTER 12

1 Richard H. Thaler and Cass R. Sunstein, *Nudge: Improving Decisions About Health, Wealth, and Happiness* (New York: Penguin, 2009).

2 Nick Toman, Brent Adamson, and Cristina Gomez, "The New Sales Imperative," *Harvard Business Review*, March–April 2017, https://hbr.org/2017/03/the-new-sales-imperative.

CHAPTER 13

1   Richard E. Boyatzis et al., "Emotional and Social Competency Inventory (ESCI): Technical Manual," Korn Ferry Hay Group, 2015, https://www.kornferry.com/content/dam/kornferry/docs/article-migration/ESCI_Technical_Manual_nav_04052017.pdf.

2   Daniel Goleman, *Emotional Intelligence: Why It Can Matter More Than IQ* (New York: Bantam Books, 1995).

3   World Economic Forum, "The Future of Jobs Report 2020," https://www.weforum.org/reports/the-future-of-jobs-report-2020.

4   Shelley Thompkins, PhD, "Emotional Intelligence and Leadership Effectiveness: Bringing Out the Best," Center for Creative Leadership, August 28, 2023, https://www.ccl.org/articles/leading-effectively-articles/emotional-intelligence-leadership-effectiveness/.

5   Angel E. Navidad, "Stanford Marshmallow Test Experiment," *Simple Psychology*, September 7, 2023, https://www.simplypsychology.org/marshmallow-test.html.

CHAPTER 14

1   Elizabeth M. Renieris et al, "Artificial Intelligence Disclosures Are Key to Customer Trust," MIT Sloan, https://sloanreview.mit.edu/article/artificial-intelligence-disclosures-are-key-to-customer-trust/.

CHAPTER 15

1   Laura LaBerge et al., "How COVID-19 Has Pushed Companies over the Technology Tipping Point—And Transformed Business Forever," McKinsey & Company, October 2020, https://www.mckinsey.com/capabilities/strategy-and-corporate-finance/our-insights/how-covid-19-has-pushed-companies-over-the-technology-tipping-point-and-transformed-business-forever.

2   Kelly Blum, "Press Release: Gartner Says 80 Percent of B2B Sales Interactions Between Suppliers and Buyers Will Occur in Digital Channels by 2025," Gartner, September 15, 2020, https://www.gartner.com/en/newsroom/press-releases/2020-09-15-gartner-says-80--of-b2b-sales-interactions-between-su.

3   Rob Zack, "Face-to-Face vs. Screens: What Neuroscience Reveals About Real Connection," MicroAge, https://microage.com/culture/face-to-face-vs-screens-what-neuroscience-reveals-about-real-connection/.

4   Nidhi Arora et al., "The Value of Getting Personalization Right—Or Wrong—Is Multiplying," McKinsey & Company, November 12, 2021, https://www.mckinsey.com/capabilities/growth-marketing-and-sales/our-insights/the-value-of-getting-personalization-right-or-wrong-is-multiplying.

5   John Suler, "The Online Disinhibition Effect." CyberPsychology & Behavior 7, no. 3 (2004), https://www.liebertpub.com/doi/10.1089/1094931041291295.

6   "Types of Social Proof: Fourteen Examples Showing Their Impact," Fomo, https://fomo.com/blog/social-proof-types.

ABOUT THE AUTHORS

Jeff Bloomfield is a leading expert in neuroscience-based communication and the founder of Braintrust. As the creator of the NeuroSelling® and co-creator of NeuroCoaching® programs, Jeff's research and techniques empower leaders, coaches, and sales professionals to build trust and inspire action through brain science, behavioral psychology, and storytelling. His mission is to help people communicate with greater clarity, authenticity, and influence—in both business and life.

Dan Docherty, PhD, is chief coaching officer and managing partner of the Leadership Development Practice at Braintrust, leadership professor in the MBA program, and assistant director of the Isaac & Oxley Center for Business Leadership at Miami University. With a PhD in Management from Case Western Reserve University, his research centered on the neuroscience of coaching, development, and performance in leader-team member relationships. A former life sciences executive, Dr. Dan has coached thousands of leaders across the globe and co-created the NeuroCoaching® framework to assist leaders in communicating with impact to build stronger relationships and drive performance.

For more information, visit us at www.braintrustgrowth.com

For consulting inquiries to have NeuroSelling brought to your organization: support@braintrustgrowth.com

Interested in having one of us speak at your next event? Please visit: www.jeffbloomfield.com and www.dandocherty.com

www.ingramcontent.com/pod-product-compliance
Lightning Source LLC
Chambersburg PA
CBHW022112210326
41597CB00047B/209